NEITHER BEAST NOR GOD

Gilbert Meilaender

Neither Beast nor God

The Dignity of the Human Person

New Atlantis Books
Encounter Books · New York · London

First American edition published in 2009 by Encounter Books,
an activity of Encounter for Culture and Education, Inc.,
a nonprofit, tax exempt corporation.
Encounter Books website address: www.encounterbooks.com

Manufactured in the United States and printed on acid-free paper.
The paper used in this publication meets the minimum requirements of
ANSI/NISO Z39.48–1992 (R 1997) (*Permanence of Paper*).

FIRST AMERICAN EDITION

LIBRARY OF CONGRESS CATALOGING-IN-PUBLICATION DATA

Meilaender, Gilbert, 1946–
Neither beast nor God : the dignity of the human person / by Gilbert Meilaender.
p. cm. — (New Atlantis books)
Includes bibliographical references and index.
ISBN-13: 978-1-59403-257-8 (hardcover : alk. paper)
ISBN-10: 1-59403-257-2 (hardcover : alk. paper)
1. Philosophical anthropology. 2. Dignity. 3. Respect for persons. I. Title.
BD450.M376 2009
128—dc22
2009020761

10 9 8 7 6 5 4 3 2 1

CONTENTS

TO NICOLAS

"*I regarded myself preferably as a* reader *of the books, not as the* author."

Preface

THIS SMALL BOOK on human dignity began to form itself in my mind as I puzzled over questions that arose in the work of the President's Council on Bioethics. Much of the book, therefore, concerns itself with bioethical matters, though it is hardly a typical work in that field.

Appeals to dignity have become very common in bioethics, though by no means only there. The word has been around for a long time, but it means many different things. The longer I puzzled over it the more I began to think that we need to distinguish especially two different senses—what I here call human dignity and personal dignity. The concept of human dignity is simply a placeholder for what is thought to be characteristically human—and to be honored and upheld because it is human. The concept of personal dignity is needed to make clear that, however different we may be in the degree to which we possess some of the characteristically human capacities, we are equal persons whose comparative "worth" cannot and ought not be assessed. I'm sure I do not have all the puzzles raised by the language of dignity perfectly sorted out here, but I hope at least to have found a helpful way into the discussion.

Much of the impetus that moved me to think through these

matters has, as I noted, come along the way in the work of the President's Council on Bioethics. Beyond a shadow of a doubt, many of my colleagues on that Council would not share all the views I articulate here, but I thank them nonetheless—and thank, at least as much, members of the Council's staff, especially Eric Cohen, Yuval Levin, and Alan Rubenstein—for the thinking they have forced me to do. In particular, I owe thanks to Leon Kass, with whom I have had many friendly disagreements as we tried to sort out the idea of dignity. He will not be completely content with what I write here, but he cannot entirely avoid responsibility for having helped to awaken—and even to nurture—these thoughts in me.

This is not a work of theology in any technical sense, but it is, in certain respects, a piece of religious thinking. That is, I doubt whether we can understand dignity well without at least a modest anthropology—without some notion of what it means to be the sort of creature a human being is. And I, at least, do not think this understanding can possibly be right if we abstract the human beings we seek to understand from their relation to God. Abstracted from that relation, they are simply abstract—not really what human beings are. Hence, I have not hesitated to think in religious terms when it seemed necessary to me, and I have proceeded from the venerable premise that faith seeks understanding. I hope to have found at least a bit of it.

CHAPTER ONE

Speaking of Dignity

Offering his own explanation for the confusing mixture of works he had written—some published pseudonymously, others under his own name—Kierkegaard distinguished between two quite different senses in which his writing had focused on "the single individual." In the works written under different pseudonyms, the individual whom he had in mind was a "distinguished person"—distinguished, that is, by human excellence of one sort or another. But in the writings published under his own name, which he called "edifying" works, the individual was "what every man is or can be." Hence, "'the single individual' can mean the one and only, and 'the single individual' can mean every man."[1]

From one perspective individual human beings are members of a species that is distinguished from other species by certain characteristics. The species lives on, though its individual members die (and, indeed, probably must die for the health of the species). Almost inevitably, some individual members of the species display more fully or more excellently than others its distinguishing characteristics. In so doing, they give us some sense of what a human being at his best can be, and we may sometimes speak of their conduct as dignified. They are distinguished from the rest of us and offer an image of the flourishing of our full humanity. In so flourishing they display what I will call *human* dignity.

This way of thinking about dignity invites us to attend to at least two different but significant matters. The first grows out of the fact that *human* dignity is the dignity of a particular sort of creature, who is neither the "highest" nor the "lowest" sort of creature we can imagine. Indeed, the term *dignity* here is really just a placeholder, a shorthand expression for a certain vision of the human. Human beings are strange, "in-between" sorts of creatures—lower than the gods, higher than the beasts. Not simply body, but also not simply mind or spirit; rather, the place where body and spirit meet and are united (and reconciled?) in the life of each person. Thus, Augustine writes, in a sentence that succinctly captures the point, God "created man's nature as a kind of mean between angels and beasts."[2]

This characteristically human dignity, this in-between status, does not always satisfy us, however, and so we may seek to be either more or less than human. As Filostrato, a physiology professor, says in C. S. Lewis's fantasy, *That Hideous Strength*: "What are the things that most offend the dignity of man? Birth and breeding and death."[3] Unsatisfied with our condition, we hope to reshape and master these central aspects of life and become more than human. Death must be conquered—or put off as long as possible. Reproduction must become the work not of the body but of will and mind—and need no longer involve animal copulation.

The motives underlying such attempted mastery need not be obviously evil; indeed, they will almost surely include a desire to relieve the pains and disappointments of the human condition. But the price paid is what Lewis elsewhere called an "abolition of man," a subverting of the character of our in-between status as beings marked by "not just reason or will, not just strength or beauty," but by "integrated powers of body, mind, and soul."[4]

Filostrato was not mistaken to suppose that competing visions of human dignity come most clearly into focus when we think about birth, breeding, and death. How we come into being and how we go out of being are of central importance for any sense of what it means to respect (or undermine) human dignity. But human dignity also involves more than how we are born and how we die. To be born of human parents is to be connected in particular ways. We are located; we are not just free-floating spirits or citizens of the world. We do not spring up like mushrooms from the ground, and we therefore have special attachments to some, even obligations to which we never consented and which we never chose. These special attachments, loyalties, and obligations are part of what it means to be a human being. They too are integral to our dignity in the time we are given between birth and death—a time marked, usually, by growth and achievement, but also, usually, by failure and loss.

When dealing with either birth or death, our greatest temptation may be to use our powers of mind and soul to control and master our bodies—to be more than human. In much of life that comes *between* birth and death, however, we are increasingly tempted to see our problems not as invitations to mindful mastery but as bodily problems to be medicated away—as if we were less than human. Life's difficulties become not an occasion for development of character and virtue but "medicalized" problems calling for a prescription.

Thus, in different ways we may think of ourselves not so much as the peculiar in-between creature in whom nature and spirit meet but as either "just body" or "just spirit." Neither of these is bad in itself. An animal is not a bad thing, nor is an angel or a god. But we are none of these, and human dignity is to be found in the kind of life that honors and upholds the peculiar nature that is ours, even if there is no recipe book that can always show us how properly to unite and reconcile body and spirit. Much of our insight into that nature is the fruit of our Christian and Jewish traditions; yet, what faith seeks and sometimes finds is insight into a true humanism—into the meaning of human dignity. Human life is marked by characteristic powers and capacities, but also by characteristic limits and, even, weaknesses. We need to honor and uphold that peculiar, in-between character of human life.

A second, related but different, issue raised for us by the idea of *human* dignity is all too apparent. Because our life is marked by characteristic powers and capacities, we are naturally inclined to think in terms of comparative degrees of human distinction or dignity—and of some as more dignified than others. Just as some of us flourish, displaying humanity at its fullest and best, others of us have, at most, a kind of basic humanity that falls far short of the full development of human possibilities—"an anthropological 'minor league.'"[5] And, in fact, it is not hard to imagine that some of us might so lack or lose the characteristic human capacities as to seem to have lost human dignity almost entirely. Thus, Peter Singer wonders why we should affirm the dignity of all human beings, "including those whose mental age will never exceed that of an infant," when "we don't attribute dignity to dogs or cats, though they clearly operate at a more advanced mental level than human infants."[6] If dignity is a comparative concept, grounded in certain capacities, which may or may not be present, some human beings will have greater dignity than others, some will conduct themselves in a more dignified manner than others, and some may have lives utterly devoid of human dignity. That is how we think if we think of an individual as "the one and only"—as distinguished by certain characteristics (or, of course, in a negative mode, by the lack of those characteristics).

Kierkegaard thought of the individual not only as "the one and only," however, but also as "every man." At issue here is not *human* dignity but what I will call *personal* dignity. The equal dignity or worth of the individual person has, in our history, been grounded not in any particular characteristics but in the belief that every person is equidistant from Eternity—and that, as Kierkegaard says, "eternity . . . never counts."[7] The God-relation individualizes. When all are equally near (or far) from God, all other distinctions are radically relativized, and one can even say that "all comparison injures."[8]

But we cannot simply set human and personal dignity side by side and say no more. For if we think primarily in terms of human dignity we may be tempted to suppose that the equality embedded in the concept of personal dignity is a fiction (if perhaps a useful one). It may seem that some of us—those whose capacities are less developed or, if once developed, are now fading and diminished—have a lesser status than others in whom the most characteristic human qualities are more fully displayed. Some people's lives—those in a persistent vegetative state, those suffering from severe dementia, those who are profoundly retarded, those who have achieved little and whose lives display no characteristically human excellences, those whose cultural achievements are few, those who have done evil deeds and show no remorse—will lack the dignity that characterizes genuinely human life and perhaps even no longer be "worth" preserving. Thinking only of human dignity, and of the unequal ways in which it is displayed in our lives, we may turn naturally to the quantitative language of "value" and conclude that the lives of some people are "worth" less than the lives of others.

If, on the other hand, we think first in terms of personal dignity, we are likely to emphasize individual equality, affirming—or insisting —that every person, simply by virtue of his or her humanity, is one whose dignity calls for our respect. Nothing we do or suffer can deprive us of the dignity that belongs to each person. We may offend against that dignity or fail to recognize it, but we cannot destroy it or blot it out. To think this way is to honor and uphold not simply a true humanism, but also a true personalism.

Still, we must, as I noted, do more than simply set human and personal dignity side by side. One of them must, at least to some degree, have a transforming effect upon the other. I suspect we can see such a transforming effect in the following example: Discussing the topic of murder, and replying to an "objection" (as the structure of the *Summa* calls for such replies), St. Thomas Aquinas writes, "A man who sins deviates from the rational order, and so loses his human dignity [*dignitate humana*]. . . . To that extent, then, he lapses into the subjection of the beasts."[9] Contrast this with the words of Pope John Paul II in the encyclical letter *Evangelium Vitæ*, released in 1995: "Not even a murderer loses his personal dignity [*dignitate*]."[10]

The seeming divergence between these two important and influential statements within the same (albeit long and extended) tradition of thought is striking. Aquinas seems to think that the murderer, by turning against what reason requires of us, becomes more beast than man—losing the dignity that characterizes human beings, the rational species. John Paul II, in a context discussing the death penalty in general and Cain's murder of Abel in particular, does not seem to think of "dignity" as something that can be lost by human beings, even when they act in ways that fall far short of the excellences that mark human nature. The divergence is, I think, only "seeming," however, for what we see here is the way in which this tradition of thought has regarded *personal* dignity as fundamental and allowed it to transform to some extent our understanding of *human* dignity. For the most fundamental truth of all is that, as Kierkegaard noted, "eternity . . . never counts."

We have, then, two concepts of dignity—human and personal—that invite our reflection. *Human* dignity has to do with the powers and the limits characteristic of our species—a species marked by the integrated functioning of body and spirit. We may differ, individually, in the way or the degree to which we manifest those characteristics and that distinctively human wholeness, but the specific dignity of the human species would be diminished or lost if we were utterly to transcend the limits of our bodies (and become something more like a god) or if we were to think of our bodies not as the place of personal presence but (as for beasts) things to be manipulated for purposes entirely external to them. The first of these ways of subverting our human dignity has been called, in our moral tradition, *pride*; the second *sloth*.

Personal dignity, by contrast, has to do not with species-specific powers and limits but with the individual person, whose dignity calls for our respect whatever his or her powers and limits may be. It is closely tied to our affirmation of human equality. Each concept is needed and merits our attention, but it is personal dignity that provides a *cantus firmus* underlying and sustaining the whole.

CHAPTER TWO

Being Human

NEAR THE BEGINNING of his *Nicomachean Ethics*, as Aristotle probes the question of what is "good" for—that is, what will fulfill—a human being, he suggests that the answer lies in "whatever is his proper function" (more literally translated, "proper work").[1] That distinctive function cannot, he says, be simply "living" (i.e., nourishment and growth), for that human beings share even with plants. It cannot be merely sense perception, which is shared with the other animals. The distinctive work of a human being, he concludes, must be to live as a reasoning being ("actions performed in conjunction with the rational element" of the soul).[2]

That is all we need to get us started in this chapter's discussion of "being human," but it is important to note one more thing Aristotle says, for it relates to complications to be taken up much later. If the distinctive work that characterizes the human species is, as Aristotle puts it, activity of the soul in conformity with reason, it will, of course, be true that some of us more fully and excellently carry out that work than do others of us. Thus, Aristotle concludes that "the good of man is an activity of the soul in conformity with excellence or virtue, and if there are several virtues, in conformity with the best and most complete."[3] The dignity of being human takes its rise from the work that distinguishes our species; yet, some of us express or manifest that characteristic work more fully than do others of us. How to affirm or hold onto the equality of human persons in the face of this difference will occupy us later.

We need to start, though, roughly where Aristotle did—with "man" himself and the work that characterizes and distinguishes the human species. In articles collected in both *The Phenomenon of Life* and *Philosophical Essays*, Hans Jonas explored and developed a vision of what it means to be a living organism and, more particularly, a living human being.[4] Beginning with a few aspects of his "philosophical biology" will give us an angle from which to think about the dignity of being human.

THE PHENOMENON

Unlike inert matter—dirt, or a rock, for example—which simply is what it is, living things retain their individual existence over time only by *not* remaining what they are, only by "special goings-on" which "efface almost entirely [their] material identity through time."[5] These special goings-on we call metabolism. Organisms remain alive by taking in material from their surrounding environment and using it to produce energy and to fashion materials for growth and repair (and then, of course, by returning to the environment the material they cannot use for these purposes). The very *being* of the organism consists in sustaining itself by carrying on this work of exchange with its environment; for without such exchanges it cannot continue to "be."

Thus, the living form, whether of the human body or of less intricately developed organisms, must be distinguished from the materials that constitute it. An organism is "a substantial entity" which nonetheless enjoys "a sort of *freedom* with respect to its own substance, an independence from that same matter of which it nonetheless wholly consists."[6] However mysterious this may seem, it is unavoidable.

"Inwardness" is built into the nature of living organisms.[7] If we were simply to identify the organism with the materials that compose it at any given moment, it could have no continuing identity over time. For when we observe it at different times we see that it has lost many of its components and gained others not previously part of it—eventually having completely replaced its previous material components, while still remaining the organism it was at our earlier observations. Indeed, if we thought in terms of the material components that constitute it, we would say that it can remain what it is only if it dies, only if it ceases to engage in the exchanges that constitute metabolism. A living being "is never the same materially and yet persists as its same self, *by* not remaining the same matter. Once it really becomes the same with the sameness of its material contents—if any two 'time slices' of it become, as to their individual contents, identical with each other and with the slices between them—it ceases to live."[8]

One might, of course, say that this is also true of a complex object such as a ship. All its parts could be replaced over time; yet we would still consider it the same ship. But its identity over time and through change is, as Jonas notes, quite different from the living organism's. Changes in the ship are carried out by external agents for their own purposes; changes within the organism are its own work in relation to its environment, carried out for the purpose of sustaining itself. More important still, whereas changes in the ship are needed only because of the accidents of damage or decay, changes in the organism are a necessity if it is to continue to be. "And whereas it would be considered a mark of excellence for the ship to withstand use without the need for repair, for the organism the mark of excellence is not immunity to wear, but efficiency of repair."[9] And, finally, it is we who ascribe to the ship its identity over time. But the organism's identity is its own achievement, as it works to maintain its existence over against the surrounding world.

* * *

PRESERVATION AND PURPOSE

In the phenomenon of metabolism, therefore, we see the germ of deep truths about living beings. There is a kind of centeredness to an organism's life. It lives only by working to sustain its life through constant exchanges with the surrounding world. (It is striking how nicely Jonas's philosophical biology coheres at this point with Aquinas's view that the first of the natural inclinations, upon which the natural law is based, is that "every substance seeks the preservation of its own being."[10]) Hence, organisms are centered, their inner identity set over against and distinguished from the external world.

Simultaneously, however, at the core of an organism's centered identity it is of necessity open to the surrounding world—with which, in order to live, it must engage in a constant series of exchanges. Its freedom from the external world is a "needful freedom."[11] For without openness to the world it cannot sustain itself. We should underscore this fact: A living being—centered as it is on the task of sustaining its own existence—cannot accomplish this if it is entirely *self*-centered. To preserve its centered identity it must risk it, must turn outside itself—a kind of "death" that is the secret of its life.

The relation of some organisms—plants, for example—to their surrounding environment, though remarkable in its own way, is comparatively uncomplicated. "Rooted" in their environment, plants simply draw the materials they need for growth and repair from that immediately surrounding environment. There is little gap between what they need to survive and what their surroundings provide. Not so for animals, whose life is, Jonas notes, distinguished from plant life by *perception* (of what is not immediately present), *emotion* (in particular, desire for what is perceived), and *movement* (to secure what is needed and desired to sustain life). Thus, an animal's relation to the world is less immediate than a plant's. It is distanced from its world by these three "modes of mediacy," which express a "split between self and world—a qualitative widening of the split which metabolism opened first, and which is thus at the root of life."[12] And for human beings this mediated relation between self and world becomes marked also by reflection—that is, the ability to produce representations of objects not immediately present and reflect upon them.

Moreover, *purpose* is embedded at the heart of a living being's needy freedom; it is objectively there, embedded in the very nature of things. To live is to engage purposively in the exchanges of matter that metabolism involves; to die is to be unable to sustain the purpose that lies at the center of the organism's identity. Jonas contrasts the way in which a target-seeking torpedo *aims* at a *goal* and might *miss* that goal with the way living beings *aim* at sustaining their existence. The torpedo has no goal of its own. If it misses its goal, what this means is that it has missed "*our* goal, the goal for which it has been designed."[13] It "carries out" a purpose, but we should not confuse that with the quite different undertaking of "having a purpose."[14] Living beings, which have a centered identity distinguished from the world external to them, and in whose very being is embedded a distinction between inner identity and outer material components, bear witness that the world is in truth teleologically ordered. "Appetite or desire, not DNA, is the deepest principle of life."[15]

PROCREATION AND SACRIFICE

A mediated relation to the world means, in turn, greater individuality. The more distinguished from and discontinuous with its environment an organism becomes, the more individual it is. And the greater the individuality, the greater also the risks involved in sustaining one's existence. What higher animals—and, certainly, human beings—need to sustain life is not "automatically" provided, as it is for plants. Because animals are not rooted in their surroundings, because there is "distance" (both in space and time) between what they need to live and the purposive actions they take to meet those needs, the work of self-preservation becomes more hazardous. Anxiety, suffering, and fear are all now inseparable from enjoyment. Indeed, for the freest and most individuated of beings—human beings—the individual's "possible annihilation becomes an object of dread just as its possible satisfactions are objects of desire."[16]

Although Jonas's analysis of the mystery of organic life directs our attention in this way to the fact of death, it does not as obviously invite us to think about birth. In an essay devoted largely to explication and appreciation of Jonas's philosophical biology, Leon Kass noted that missing from it was an account of "the specially focused kind of desire which is rooted in sexual difference."[17] What we need to add, Kass suggests, is a more complicated account of human desire (one of those modes of mediacy). Central to the living being's desires are, of course, desires closely related to the task of preserving its life—a hunger to eat, for example. But animals such as human beings, who reproduce sexually, also experience a desire to mate (and, hence, the second of the natural inclinations from which, according to Aquinas, the natural law derives, is the inclination to engage in sexual intercourse).[18]

This is, in the first instance, a desire not for motherhood or fatherhood but for that one who though like us is also other than and complementary to us. Thus, sexual desire—an *eros* that is more than mere appetite—directs us to the most fundamental form of human community, the nobility of which lies in part in the fact that here the organism whose work consists in maintaining itself is drawn into a bond which might seem to need no justification beyond itself. Thus, noting the (to some) curious inclusion of the Song of Songs in the Old Testament canon, Karl Barth wrote: "We should not wish that this book were not in the Canon. . . . And we should not spiritualise it. . . . [I]t is a collection of genuine love-songs in the primitive sense, in which there is no reference to the child, but only to man and woman in their differentiation and connexion, in their being in encounter."[19]

Yet, the *telos* of this desire, as Kass notes, goes well beyond the immediate object of passion. Its end is that we produce children who will replace us. There is irony embedded in nature's purpose here. The human beings whose central work as organisms is the ongoing task of sustaining their own existence through metabolism are also moved by a desire that *first* draws them into a bond of love which, though seemingly unnecessary to sustain their life, embodies the needy openness that Jonas regarded as so characteristic of living organisms—and *then* produces those who will take their place. We had noted earlier how Jonas's explication of organic life demonstrates that living beings can sustain their existence only if they also take the risk of turning outward to their world, only if they are not simply self-centered. Kass's extension of Jonas's depiction of animal life—and, in particular, human life—presses this point even further: "In sex, life is *not* just self-centered individuality, on the contrary, sexual desire, in its deepest meaning, is self-sacrificing."[20]

LOVE AND WORSHIP

In this brief depiction of what it means to be a living human being, we can discern a good deal about the aspects of our lives for which the language of "human dignity" serves as a placeholder. The ability to discern this significance is not a given, however. Because the life of human beings, though embodied, has an inner dimension, it will always remain somewhat mysterious. And it will always be possible to see only a part of the truth—reducing our humanity to body alone or to free spirit alone. In so doing, however, we fail to honor the peculiarly "in-between" character of human beings, who are neither beasts nor gods.

Human life must be marked by a certain duality: As material organisms we work to live as all organisms must, but with an inner freedom from identification with our own substance, a freedom that is the ground of desire and purpose. But if we permit this *duality* to become a *dualism*, we separate these two aspects of our being instead of distinguishing them in their connectedness, and we begin then to think of human beings as—to use the old language—either just body or just soul. Quite often today, in fact, these two seemingly different reductionisms may coexist in our thinking. We suppose that we are free to reshape ourselves without limit (as if we were just soul), *and* that reshaping takes the form of medicating ourselves (as if we were just body). Avoiding the temptations of such dualism is necessary if we are to honor human dignity.

That dignity, we can now see, involves how we come into being and go out of being, each of which carries deep meaning for the "in-between" creature we call human, but it involves more than just those beginning and ending points of life. When the sexual act gives rise to offspring, it creates—at least among human beings—kinship bonds that mark out those to whom we have a special relation and for whom we incur special obligations. In recognizing these obligations, we discover not something we have invented but something deeply embedded in the nature of things. As organisms characterized by "needful freedom," a kind of inner independence from the material of which we simultaneously consist, we find ourselves connected and obligated—needy—from the very start.

On the one hand, each self is centered. Each is engaged in the ongoing work of preserving his own life in a world where continued existence—because it is, in fact, "work"—is never guaranteed. But, on the other hand, the very project of preserving one's life requires a turn outward to the surrounding world. To exist as individuals we must not try to be *only* individuals. And in distinctively human life this simple biological truth is taken up into something higher, displaying itself in the most central aspects of our lives. As C. S. Lewis put it:

> In love we escape from our self into one other. In the moral sphere, every act of justice or charity involves putting ourselves in the other person's place and thus transcending our own competitive particularity. In coming to understand anything we are rejecting the facts as they are for us in favour of the facts as they are. The primary impulse of each is to maintain and aggrandise himself. The secondary impulse is to go out of the self, to correct its provincialism and heal its loneliness. In love, in virtue, in the pursuit of knowledge, and in the reception of the arts, we

> are doing this. Obviously this process can be described either as an enlargement or as a temporary annihilation of the self. But that is an old paradox; "he that loseth his life shall save it".[21]

For us, therefore, to exist—to "be"—is an invitation to honor the meaning of our birth and our death, an invitation to discern the centrality of obligation and love in our lives. The centered self, whose fundamental "work" is his own survival, is drawn out of himself and invited to enter into the mystery of sacrifice. How we live, not how long, is at the heart of human dignity.

Perhaps we can even extend Jonas's philosophical biology one step further and note that the *eros* which draws us toward one who is like us in being human yet other than us (because of the opposite sex) does not reach its final end when the generation that will succeed us has been born. The continuation of the species, though of great significance, is only a partial answer to the dread of death Jonas discerned embedded in human existence. For the individual is not the species, and the desire that moves us is a desire that cannot rest entirely content in a mate and offspring—as if the death of each individual mattered not at all, or as if any human being could simply be replaced by those who come after.[22] The *eros* that draws us to marry and have children is an intimation of a desire for a satisfaction still more enduring. It is a desire for God, in relation to Whom each of us is honored equally—and, unsurprisingly, the third inclination from which Aquinas derives the principles of natural law is the inclination to know the truth about God.[23]

If, as we noted earlier, purpose is embedded in all organic life, including human life, then we cannot understand the meaning of our humanity only in terms of our biological origins. To understand those origins is to see that we must characterize our humanity not only in terms of its nature, but also in terms of its destiny. If, with our unaided powers of reason, we examine the character of the organic life human beings share, that destiny can only point to our mortality and the defeat of desire. Sooner or later the fires of metabolism lose their intensity and, eventually, go out. Human life may continue, but this particular person does not—or so it must seem.

What an analysis of organic life taken alone does not enable us to see or say, however, a faith that seeks understanding may affirm. Human beings have been created for covenant community with God. And to say, as Christians do, that "Christ is risen" and "Jesus is Lord" is to believe that God is unwilling that the promise of this community should be defeated. The hint of such possibilities, embedded in the purposive character of organic life, is taken up and perfected in that life of the Spirit which overcomes death. That life is already present in —what would be unintelligible had it not happened—the risen body of Jesus, which precisely as body is no longer subject to death. In that life the fires never burn out, the hints buried in organic life become realities, and desire is satisfied.

Thinking about human dignity, about our needy openness to the world around and beyond us, reminds us that our humanity cannot adequately be described apart from the relation to God, and this—taken seriously—must eventually press us to think not only about our shared human dignity but also about the dignity of each person.

Human Dignity

CHAPTER THREE

Birth and Breeding

We have begun recently to speak of "transhumanists," people who look forward to a world in which aging has been overcome and our physical and intellectual powers have been enhanced. They suppose (and hope) that we may produce beings who are "posthuman," who—at least measured by the standard of our own powers and capacities—seem more than human. Characterizing certain well-known thinkers[1] as opponents of such transhumanist imaginings, Nicholas Agar depicts their view as follows: "Although there are differences between them, these thinkers share a desire to keep us and our near descendants human, even if this means keeping us and them dumb, diseased, and short-lived. They identify the technologies that enthuse transhumanists as distinctively threatening to our humanity."[2]

This may not be the fairest (it is certainly not the most appealing) way to characterize the opponents' position, but it points us to something important. If we destroy what is distinctively human about ourselves in exchange for producing persons whose condition (with respect to suffering, death, and various capacities) might be described as "better" or more advanced than ours, what does this mean for human dignity? Even while ascribing to these new—higher?—beings the dignity that belongs to persons, we might sensibly also say that our human dignity had been demeaned or lost. And we might, were we at all sympathetic to the "opponents" of transhumanist desire, wonder whether it would have been better to remain human—characterized by a needy openness that exists only by virtue of constant exchanges with a world we do not master—even if our capacities were fewer, our status (in some sense) lower, and our suffering greater.

Why should we want to be or remain human? A life free of pain, without death (if we can imagine that for ourselves), with greatly enhanced powers (of memory, for example)—a life more godlike than human[3]—might seem preferable. Likewise, a life in which unhappiness or a sense of failure could simply be medicated away could easily, at least in certain moods, have its attractions for us; yet it would be a life in some respects "lower" than ours, as Mill said it would be worse to be a pig satisfied than Socrates dissatisfied. Why want to be human, this strange creature who can experience neither the uncomplicated wholeness of a beast nor the mastery of a god?

GRATITUDE

Perhaps our first concern should be to kindle or rekindle in ourselves a sense of wonder at our humanity, situated so precariously at the juncture of nature and spirit, body and soul. We ought not lose the capacity for gratitude. It is good that there should be creatures such as we are. Unless and until we can say that, we are condemned to an endless attempt to be something else—whether that something else seems to be "less" or "more."

Attachment to our own species and its welfare is suspect in the eyes of some. "What magic is there in the pronoun 'my,'" William Godwin famously asked, questioning the rightness of special concern for ourselves, for those closest to us (and by extension, one might suppose, for our species).[4] The very question displays a desire to be something other than we are, something more than human; for it seeks a God's-eye view, a view located nowhere in particular.

Actually, it is not that hard to answer Godwin's question. We need only consider what would be lost if we really tried to live as citizens of the world (or universe), or as minds merely using or inhabiting our bodies but not located and limited by the body as the place of personal presence. What would be lost?

My home—local and unique, a particular place and time. *My* child —bound to me by biology and history in ways he or she is bound to no other. *My* accomplishments—whether few or many. *My* memories —a tissue of connections that form my personal history. *My* failures and, even, sins—which, however bad in themselves, provide the occasion for something good: shame and renewal. *My* attachments and loves—which give a pleasure embedded in the relationship itself and unavailable apart from such attachment. *My* friends—bound to me by both the bodily tie of close proximity and the spiritual tie of shared concerns. *My* generosity toward strangers—which could not be generosity were it simply my obligation toward every human being. *My* vulnerability to loss—inseparable from my loves. *My* suffering—which, though I would gladly be relieved of it, provides an occasion for the human virtues of courage, patience, and hope. For all this the language of human dignity may serve as a simple placeholder, and when we use the placeholder we express both our wonder at the goodness of such a life and our determination to honor and uphold its distinctive character.

That distinctive character begins precisely with the magic that is embedded—and rightly so—in the pronoun "my." A characteristically human life is marked from the outset by special moral relations, by bonds that are personal and private. They are personal—in the sense that those who share this bond cannot be interchanged with any others; they are private—in the sense that their boundaries are limited to some and others are excluded.

We can imagine a life not characterized by such bonds. There could, for example, be creatures who regarded each other as entirely interchangeable. They might cooperate or join together for certain purposes, but they would not form *personal* bonds, in which one individual could not without loss be interchanged with another of the species. We could also, perhaps more interestingly, imagine a Being Who, while loving all others personally (that is, loving each person in all his or her particularity and as non-interchangeable with any other), would not love in a way that was private (that is, would exclude no one from the scope of His love). Either of these can be imagined, and the second—a love which, though universal in scope and excluding no one, cares for each person with the particularity and passion that we experience only in our most intimate loves—is a love so admirable that we think of it as the way God loves. Admirable as it is, however, one might be almost reluctant to characterize it as human—not because it is inhumane or less than human, but because it is so much higher than any love of which we seem naturally capable.

That we can imagine a life higher than human life does not mean we should draw back from honoring and upholding the dignity of our peculiarly human condition. Indeed, it is part of being human that we should know ourselves to be something other and higher than simply animals—but that this otherness should consist especially in the longing or desire for what is still higher and better than we, the longing for God. Human existence is lived "in between," and to honor human dignity is to honor precisely the in-betweenness that marks our condition.

The human person—neither beast nor god—is a real union of body (that ties us to the beasts) and soul (that directs us toward God). In us, inwardness is inextricably tied to organic life. When, however, we try to articulate what this means (especially in religious terms), we may picture the human person as a composite of two things that are in principle separable, that are temporarily glued together in this life, that will (by God's grace) be separated in such a way that the person continues to live even after the body has died, and that will one day be reunited (in a resurrected life). That picture, appealing as it has been at different times and places, is more dualism than duality. It does not fully capture our in-betweenness, which is not simply a composite of two essentially different things (such as horse and rider).

We might, borrowing an image from C. S. Lewis, revise that picture which, though defective, comes so naturally to us. Instead of horse and rider, think of a centaur. In the case of horse and rider, we might shoot the horse out from under and have the rider survive unscathed. Likewise, we can imagine the living horse without the rider, just an animal free of human direction. By contrast, the centaur is a real union, the union of man and animal—as the human person is a union of body and spirit.[5]

We can, therefore, discern two general ways in which we might fail to honor our distinctively human condition. We might think of human beings as just bodies: a complicated animal, to be sure, but one for whom the animating principle is, finally, complex chemical interactions of the brain, in terms of which anxiety, fear, suffering, love, and joy can be explained. Or we might picture human beings as, really, just souls: an immaterial consciousness that is not essentially embodied and that may one day be able to cast off the biological substrate upon which it currently depends and achieve a kind of intelligent immortality that is not dependent on the body and that no longer need look out in needful freedom upon the world. The first of these might be called subhuman, the second posthuman; either might intelligibly be thought to undermine human dignity. Not because a being who is "just body" or "just soul" is a bad thing, but because it is not the human being whose existence we honor, who occupies a peculiar in-between place in the world, and who is made, so Jews and Christians believe, to be God's covenant partner in caring for the creation. It is good that human beings should exist.

MEN AS MUSHROOMS

Every human being is connected to certain others in special relationships. These bonds may be largely spiritual—as is the case, for example, in friendship. They may be deeply grounded in biology—as is the relation between parents and their children. These relationships obligate us in specific ways, and they are not just instances of or derived from more general commitments that we have to all others. Of all such special bonds, that of parents and children cuts more deeply than most into our identity, shapes our character in lasting ways, and is grounded in the sexually differentiated character of human life.

Some have tried to understand the family solely by means of the language of rights, contract, and choice. Hobbes, for example, in order to secure the Leviathan state against potentially subversive family loyalties, described the family as a "little monarchy"—just a smaller political community, and able, therefore, to be overridden by the larger Leviathan when loyalties clashed. But the language of rights cannot account fully for the family's importance in human life. A father's rights have not necessarily been violated if he is unable to feed his children. Nonetheless, his human dignity is diminished. And a decent community will do what it can to avoid this, not because the father has a right to feed his children, but because the human dignity we share is undermined when he cannot.

Or, again, a world in which adults sought to treat all children impartially—entirely oblivious to the fact that a few of those children were *their* children—would not be a fairer or more just world. Perhaps God does and should love that way, but we are not God. Such a world would be impoverished; for it would have no place where we simply belonged to others, needing to offer no justification to support our claims upon them.

This is why—even bracketing entirely more general arguments about abortion—the ready acceptance of abortion of "defective" fetuses (or, now, assisted reproduction procedures in which "defective" embryos are selected against), violates the human dignity we share. It sets aside the fundamental bond of parents and children, inserting choice in the place of love and acceptance, and teaching us thereby that we must justify our continued existence, especially when we constitute a burden to others. That is inhumane in the most precise sense, for it drains moral significance from a relationship which deeply marks our human identity and which makes space in life for a love that need not be earned.

Suppose, Hobbes writes in *De Cive*, we "consider men as if but even now sprung out of the earth, and suddenly, like mushrooms, come to full maturity, without all kind of engagement to each other."[6] This in some ways strikingly contemporary thought experiment should make clear that the dignity of the relation between parents and children was first compromised in our thought, not in our technologies of reproduction. Hobbes's concerns, of course, were political. If human beings come into existence entirely free of attachment to others, then all relations are, he thinks, essentially political. One person can have authority over another only through sovereignty by institution or sovereignty by acquisition (of which the family is an instance). It is no surprise, then, to note that the love central to the family bond—the way in which our work of self-preservation becomes self-sacrificing—is largely lost on Hobbes. He can think of no reason "why any man should desire to have children, or take the care to nourish and instruct them, if they were afterwards to have no other benefit from them, than from other men."[7]

Hobbes's human beings are all will and choice—and no body. Children, as he imagines them, are not born into any institution which corresponds to our concept of the family nor under the care of any person who is father or mother according to our traditional understanding of those roles. Indeed, there is nothing in his picture that could quite be described as a relation between the generations; for there are only sovereigns and subjects. It was with such a Hobbesian vision in mind that the philosopher A. I. Melden wrote—in language which, however commonplace it may have sounded in 1959, indicates how our technology has begun to catch up with Hobbes's supposals: "If we should live in a world in which reproduction of human-like beings were biologically unorthodox and the young were raised in queer circumstances, there need be no utility served by favoring one's forebears, in whatever strange sense this term would then be employed, but equally there would not be parenthood, family or any of the moral concepts that surround these ideas."[8] The dignity characteristic of the relation between human generations would be missing from such a world, even if sovereigns by acquisition came to love the little subjects whom they acquired.

HOBBES'S CHILDREN

Our embrace of assisted reproduction technologies—not just the use of them, but our willingness to take their availability for granted—is increasingly producing a world in which children are, like Hobbes's mushrooms, connected to us only as we will that they should be. "Hooray for Designer Babies!" Ronald Bailey entitles a chapter of his book making the "scientific and moral case for the biotech revolution."[9] If that is what the revolution means, however, we might not want to stand at its ramparts. As Paul Ramsey put it, "Men and women are embodied as well as desiring and calculating creatures." Procreation remains human, therefore, insofar as "it is not simply an activity of our rational wills."[10] Our increasing tendency to think in reflexively Hobbesian terms about birth and breeding must, therefore, have implications for human dignity.

In a general discussion of much current bioethical thinking (though, of course, primarily their own thinking) about "reproductive choice," Rebecca Bennett and John Harris begin not with any sense of what it means to be human or the place of procreation in human life but, simply, with choice. Noting the various technologies that have in different ways expanded reproductive choice, their initial question is: "What should and shouldn't constrain choice in reproduction?"[11] They begin—and, in fact, pretty much end—with autonomy, a central aspect of which is the decision whether or not to "reproduce." Human dignity consists in freedom to choose and in respect for such freedom. Thus, Bennett and Harris can say, "the principle of respect for autonomy is based on a respect for what, arguably, makes human life valuable, the ability to be the author of one's own life."[12] We should not lose our capacity to be taken aback by such a formulation, incompatible as it is with the virtue of gratitude. Clearly, a human being who is "author" of his own life (with, presumably, *author*ity over it) is not that in-between creature who, though made "little less than God," as the psalmist says, is nonetheless *decisively* less precisely because he is not the author of that status and is moved by a longing to worship what is higher.[13]

This emphasis on autonomy is not surprising and, in a sense, not even inappropriate if one thinks that human nature has no *telos*, no way of life that constitutes its particular flourishing. "One of the prerequisites for this idea of continual human self-creation," Kurt Bayertz has written, "is a farewell to teleology."[14] This is why our examination of human dignity had to begin, as it did in the previous chapter, with Hans Jonas's explication of how purpose (first as self-preservation, but then also as procreation and self-sacrifice) is embedded at the heart of an organism's—and certainly a human being's—existence. If that explication is correct, autonomy cannot be the central (or sole) characteristic of our humanity. How we come into being and preserve our existence is just as integral to the dignity of being human. If no such truths are embedded in our being, we are likely to find ourselves committed to permanent self-alteration and self-creation—no longer honored sufficiently because made in the image of God, but, instead, gods ourselves.[15]

It is not surprising to find autonomy largely preempting all other aspects of human dignity once one has said farewell to a vision of our humanity grounded in more than choice alone. So, for example, Bennett and Harris write that any limits on reproductive autonomy "must have stronger justification than mere offence or disapproval"[16]—tone deaf, it would seem, to the fact that such disapproval is grounded in a developed understanding of being human, and, still more, unable to see that their own criticism of competing views must also sound like "mere disapproval" unless they acknowledge that it too is grounded in a metaphysical account of what it means to be human.

We can explore the notion of reproductive liberty further if we consider briefly recent arguments made by John Robertson, one of the most persistent and serious advocates of a reproductive liberty that has few limits.[17] Discussing in detail the permissibility of using (our increasingly large supply of) genetic information to select and (perhaps) design our children, he distinguishes three positions, which he calls strict traditionalism, radical liberty, and modern traditionalism. Of the radical liberty position, which places almost no limits on choice in reproduction, he writes that "proponents are probably few in number."[18] Indeed, that position seems to exist in his typology primarily to permit himself to locate his own view comfortably in the center.

The chief problem Robertson finds in the strict traditionalist view, which sees more to human dignity in reproduction than autonomous choice alone, is that its "religiously based or metaphysical view of how reproduction should occur" makes it ill-suited to serve as a guide for public policy.[19] No matter how often this sort of claim is repeated by Robertson or others, however, we must remain skeptical of it. The depiction of human life developed in the previous chapter—as the highest development of organic life known to us—makes clear how mistaken we would be to characterize human existence simply in terms of our liberty to make and remake ourselves. Moreover, a view such as that defended by Robertson, which disaggregates the act of reproduction into various parts, which are then combined and recombined as we choose, is itself a "metaphysical view of how reproduction should occur." For, as Paul Ramsey once noted, it puts forward what we might call a new (and dualistic) myth of creation, according to which human beings are created with two separate capacities—the body to express the unity of the partners through sexual relations, and the power to produce children through "a cool, deliberate act of man's rational will."[20] This is every bit as much a comprehensive, metaphysical vision of the human as is the traditional view that the life-giving power to produce children should not be separated from the love-giving embrace of husband and wife. Either, or neither, may be right. But the one is not more clearly metaphysical than the other.

Reproduction, as Robertson thinks of it, is, first of all, a matter of what one hopes to accomplish. To reproduce, or to be a parent, is simply to be one who superintends and directs a project—which may involve a number of other people—intended to result in a child.[21] There are countless ways to carry out such a project, to "make" a child. These many ways do not, of course, all amount to "doing" the same thing, but that is, for the most part, unimportant for Robertson's analysis. Moreover, the liberty to carry out such a project belongs not to a couple but to an individual. It is not grounded in the characteristically human kind of desire that has its basis in sexual difference. Instead, reproductive liberty is grounded in an individual's desire to accomplish something of great personal significance, to achieve "an experience full of meaning and importance for the identity of an individual and her physical and social flourishing."[22] Such a "parent" may well, I am confident, care about and invest deeply in the child so produced. One thing such a "parent" cannot do, however: regard the child produced as made to satisfy the desires or serve the purposes of no one.

The degree to which reproductive liberty is for Robertson an individual liberty is striking; yet, more striking still is the way in which even the individual—as a human agent—begins to recede and be lost. More than once Robertson describes the "basic reproductive project" not as producing a child but as "haploid gene transmission."[23] That, of course, is clearly the project of an individual, not a couple. One produces and rears offspring "in order to transmit genes to the next and later generations."[24] Not the making of a child, but the survival of one's genes, is the project undertaken—or so it begins to seem. And, in fact, reproduction is no longer the organism's self-sacrificing turn outward, but is "behavior" driven by other forces. "A biologic perspective on human behavior suggests that reproductive success is as important an issue for humans as it is for other organisms."[25] Contrary to the vision of being human developed in the previous chapter, DNA, not desire, becomes the deepest principle of life. Having begun with a notion of reproduction governed by will and choice alone—which, however inadequate, is, at least, a truncation of the fully human integrated action of one who is body, mind, and spirit—Robertson ends with reproduction as behavior driven by entirely impersonal forces, not recognizing, it seems, how this may undermine his most basic claims about reproductive liberty. For when the point of this liberty is haploid gene transmission driven by our DNA's effort to perpetuate itself, it becomes harder and harder to know why we should be so fervent in our commitment to the notion that "reproduction is an experience full of meaning and importance for the identity of an individual."[26]

PROCREATION AND HUMAN DIGNITY

Reproductive liberty as depicted by Robertson, Bennett, and Harris disaggregates reproduction and makes it chiefly an act of will rather than the deed of one who is an integrated unity of body, mind, and spirit. The body remains necessary in certain ways, of course—as a supplier of gametes, for instance. But, no longer the place of personal presence, it is a resource used by the real person, who seemingly floats free of it. Even as relatively unsophisticated a technique as artificial insemination by donor has striking consequences for how we understand the relation between the generations. Courts may have difficulty deciding who is a child's real father or, even, whether a child has anyone rightly called his or her father.

When we turn procreation into reproduction, disaggregating its parts, we create difficulties for ourselves that we do not always want to acknowledge. Although being a father or a mother then means simply that one has commissioned the services of others in a reproductive project intended to result in a child, it may take us a while to learn to think consistently in such terms. The man who fathers a child because of a one-night stand will be held legally responsible to support that child throughout his minority. "But if a college student visits the local sperm bank twice a week for a year, produces a dozen children, and pockets thousands of dollars, he can whistle his way back to econ class, no cares, no worries." Thus, Kay Hymowitz notes, "by going to a sperm bank, women are unwittingly paying men to be exactly what they object to."[27] This is well-nigh inevitable, however, once we turn reproduction into an individual project, whether thought of as haploid gene transmission or not.

Our willingness to take such possibilities for granted means that an effort of thought and imagination is needed if we are to recapture the meaning of procreation as a fully human act. It is sad but true that the attempt to recapture the dignity of procreation, to reaggregate what has been disaggregated, itself becomes a kind of choice. That is where we find ourselves, however, and the effort must be made to reclaim an understanding central to human dignity.

When we use the language of dignity in this context, it functions largely as a placeholder for a larger vision of what it means to be human. The character of human life is degraded or diminished if we envision the relation between the generations in a way that makes some strong and others weak, in a way that makes some a "product" or an "artifact" of the will and choice of others. And this is true whether or not those who are "produced" by the will of others seem to be harmed or think themselves to have been harmed. It is true even though we need not doubt that children produced in ways which diminish our humanity will usually be loved by their parents.

We can see this if we remember how responsible parents think about their children after they are born. They do not consider them replaceable. As Ryan Anderson and Christopher Tollefsen put it:

> What parents would think, that is, that if a better version were available, they would be well advised to eliminate their present child, in order to upgrade to the better model? . . . Yet the world of assisted reproductive technology is shot through with the language of "spares," with the grading of embryos A through D, with the elimination of the lesser in favor of the better. . . . In all these ways parents manifest the depersonalizing mindset of a process that seeks to create children according to the parents' own specifications. As with any process of manufacture, refractory material is eliminated, faulty attempts are scrapped, and the drive to mastery over what is made is allowed full reign.[28]

The nature of the act—what they have *done*, not what they have *accomplished*—will have been altered significantly.

In human procreation the child is not simply a product of the will or choice of his progenitor. He is, instead, the internal fruition of an act of love. Hence, although there are different ways to *make* a child, they do not amount to *doing* the same thing; for the nature of what we do is not determined simply by what we accomplish or produce. In distinctively human procreation—an act not simply of godlike will nor animal appetite, but of that in-between creature for whom the body is the place of personal presence—the child is not a product, but a gift or a blessing.

When we disaggregate the sexual act into its parts, using the body for our purposes—which we may do, of course, with the very best of motives, hoping to help those who desire a child—the nature of the act is changed. Its result, in one sense, is not. We may accomplish the desired result and produce a child. But what we have done is quite different; it is making rather than begetting. Moreover, once we take ourselves to be producing a product, we quite naturally—even appropriately—think in terms of "quality control" of that product. All very natural, of course, but incompatible with understanding the child as the *telos*—not of our own choosing—of the work of being human.

The phrase "begotten, not made" comes from the Nicene Creed, which dates from the fourth century. It is language Christians formulated to describe Jesus as the Son of the Father—a Son Who, from eternity, is "begotten, not made." That language is intended to assert an equality of being. Christians needed a way of speaking that would enable them to distinguish the Son from the Father while yet asserting that Father and Son shared equally in the divine life and were equal in dignity. For that, the language of "making" would not suffice, since what we make, the product of our will, is not our equal. Christians therefore learned to speak of the Son as eternally *begotten* of the Father. We, by contrast, are made, by God through human begetting. Hence, although we are not God's equal, we share together the dignity of being human. We are therefore not at each other's disposal, not fit subjects for "quality control" by one another.

When Robertson discusses prenatal and preimplantation genetic screening for "non-medical selection" (i.e., for reasons of enhancement or design, rather than therapy), he considers selection of three sorts: for sex,[29] for perfect pitch, and for sexual orientation. His general view is that, if individuals or couples would not exercise their right to reproduce unless permitted to select for such characteristics, their right to reproductive liberty probably includes the right to such selection or design. Indeed, pressing still further, he even finds a "plausible case" to support cloning by those who are "gametically infertile," since they could not otherwise exercise the right to reproduce.[30] Robertson does, for the most part, draw the line at what he calls "intentional diminishment"—genetic alteration intended to produce a child lacking capacities "that would otherwise have made the child normal and healthy."[31] He allows, however, that the deaf or the dwarf "might have a plausible claim of reproductive interest."[32] (Bennett and Harris are also prepared to endorse cloning at least in cases of "infertile couples who wish to have a child who is genetically related to one or possibly both of its parents," and they seem prepared to accept ectogenesis or male pregnancy, should either become possible.[33])

Here I am less interested in debating these points than in noting how profound for our understanding of the relation between the generations is the shift from procreation as *telos* of being human to reproduction as autonomous choice. Just three decades before Robertson wrote—the blink of an eye in world-historical terms—Paul Ramsey had considered the implications of the fact that assisted reproduction technologies were less likely to treat and remedy a medical problem than to provide the desired product by other means. And, he wondered, if medicine makes this turn to "doctoring *desires*," then "is there any reason for doctors to be reluctant to accede to parents' desire to have a girl rather than a boy, blond hair rather than brown, a genius rather than a lout, a Horowitz in the family rather than a tone-deaf child, or alternatively a child who because of his idiosyncrasies would have a good career as a freak in the circus?"[34]

What ought to strike us is that Ramsey focuses on roughly the same alterations as those considered by Robertson: sex selection, perfect pitch, intentional diminishment (lacking only the example of sexual orientation). But what Ramsey sees as potentially dangerous consequences, reasons not even to start down the road of assisted reproduction, Robertson treats, for the most part, as largely acceptable consequences of our commitment to reproductive liberty. When the meaning of human dignity is narrowed to will and choice alone, little can remain of that fuller understanding of our humanity for which the body is not just a resource but the place of personal presence. The society Robertson depicts might, in a certain sense, be said to be just—scrupulous as it could be in the protection of choices that do no harm to others. But such a society, even if just, would have a diminished sense of human dignity. That dignity is something other and deeper than either justice or rights, as becomes apparent when we think with care about birth and breeding.

CHAPTER FOUR

Childhood

BIRTH, BREEDING, AND DEATH are the aspects of life most offensive to us, Filostrato suggests in *That Hideous Strength.* By this he means that they seem to mire us in our flesh, keeping us from soaring above the body and mastering it in the way he thinks appropriate to human dignity. I have argued that, properly understood, "being human" should not mean seeking to transcend the body, as if we were "just soul" or were godlike.

Although "breeding" may refer to gestating or propagating offspring, it also has a second dictionary definition as well: to nurture and train one's children. And the irony of our world is that, although we wish to exercise a kind of mastery of will in the production of these children, in the rearing of them we now often seem to act as if human beings were little more than animals—"just body."

This reduction of our full humanity is hardly better than that which, exalting will and choice, thinks of us as free spirits who transcend the body's meaning and limits. In order to examine it, we can think about one small part of life between "birth and breeding" and "death"—namely, childhood: its nature, its anxieties, its potential for growth. For if we turn childhood's problems only into occasions for seeking the right prescription, we may miss essential aspects of being human. But, in order to think about these matters, I begin very far from our contemporary concerns.

AUGUSTINE'S DISCONTENT

"You have made us for yourself, and our hearts are restless until they rest in you."[1] Thus St. Augustine, in one of the most well-known lines of one of the most influential books ever written. The *Confessions* is a story of discontent and longing, of a search to quiet that discontent, a search that—Augustine thinks—can succeed only when the heart rests in the God for Whom we are made.

The roots of his later discontent and of his impulsive and (even) self-destructive behavior were evident, Augustine believes, in the effort he made even as an infant to have his wants met and in the anger he displayed when they were not. As a young boy in school, his restless heart was evident. He preferred playing to studying and was disciplined harshly for that preference. He disliked some subjects that he was compelled to study—Greek, for example, which he never really mastered even as an adult. He was entranced by other subjects—Latin literature—and describes himself "weeping at the death of Dido."[2] He told lies to his teachers and parents, "all for love of play, eagerness to watch worthless shows, and a restless hankering to imitate them."[3]

Indeed, "in that youth of mine," he says, perhaps with the exaggerated seriousness of the convert, "I was on fire to take my fill of hell."[4] And, in an episode his recounting has made famous as a paradigm of youthful vandalism and the mystery of evil in the human heart, he and some friends stole pears from a neighboring vineyard—not because they were hungry, but because "our real pleasure was simply in doing something that was not allowed."[5] Indeed, by the time he was well into adolescence, he had become to himself "a wasteland"[6]—as he writes, in words that provided a title for T. S. Eliot's most famous poem.

Then, "I came to Carthage, and all around me in my ears were the sizzling and frying of unholy loves."[7] His restlessness and discontent, having never really found any good that answered to them in his early years, were now full grown. "I was not yet in love, but I loved the idea of love. . . . I hated security and a path without snares."[8] Despite the lack of focus and the impulsive behavior that he describes, one could not say that the maturing young man's achievements fell short of those in his peer group. In fact, he advanced professionally and became a teacher of the liberal arts. If he fell short, it was in terms of his own potential and his own desires—a failing that is painful enough, of course.

Successful as he was in some senses, however, he was far from content—and still "panted for honors, for money, for marriage."[9] He was struck by the fact that a poor, drunk beggar on the street appeared to be happy and enjoying himself, having found without trying, it seems, the contentment Augustine lacked. And when he eventually resigned his position, it was partly because, as he writes, "my lungs had begun to give way as the result of overwork in teaching. I found it difficult to breathe deeply; pains in the chest were evidence of the injury and made it impossible for me to speak loudly or for long at a time."[10] This ailment, which, so far as we know, never recurred later in his life, invites our speculation. "It is more than probable," Peter Brown, Augustine's biographer, suggests, that "Augustine had come to develop the physical manifestations of a nervous breakdown"—the end point, perhaps, of problems that had been developing throughout his life.[11]

Out of this experience of his restless heart Augustine made a great story—and, indeed, something more than that. He constructed a narrative that—in his mind and the minds of many readers since—unfolds important truths about every human life, not just his own. Moreover, although Augustine believed that only the grace of God can bring healing and wholeness to the restless human heart, he by no means absolves us of responsibility for the sad state in which we find ourselves. The fetters that bound him, as he engaged in a seemingly endless pursuit of something that would end his discontent, were not "put on me by someone else, but by the iron bondage of my own will."[12] The self is divided—longing for a contentment that only God can give, but simultaneously grasping at one or another passing pleasure—and

"I, no doubt, was on both sides."[13] That is to say, Augustine will not pretend that the discontent, the restlessness, the impulsivity, the lack of focus and readiness to be distracted are simply a misfortune he suffers—as if these characteristics were one thing and his character another. No, it is he himself who is discontented, impulsive, distractible—and to feel shame and guilt for the person he has become is a hopeful sign.

Thus, the discontent and restlessness that he experiences are not entirely bad, nor would he necessarily be better off were he completely content. For God uses this discontent to draw Augustine on toward his true happiness, which alone can offer the heart lasting peace. "You were always with me, . . . scattering the most bitter discontent over all my illicit pleasures, so that thus I might seek for pleasure in which there was no discontent and be unable to find such a thing except in you, Lord, except in you."[14] Or again: "My God and my mercy, how good you were to me in sprinkling so much bitterness over that sweetness."[15] And yet again: "Your graciousness to me was shown in the way you would not allow me to find anything sweet which was not you."[16] And even after his momentous conversion experience in the garden, Augustine knew himself to be only on the way—seeing in the distance the land of peace, but not yet there. Hence, the God Who is our happiness may not make us happy here and now. Were a pill available that seemed to make us perfectly content, we should reject it, lest we lose that eros for the God Who is truly our good. That is, we can have a false happiness, which cannot ultimately satisfy, or we can relinquish the desire for perfect contentment here and now, leaving open in our being a gaping wound that God must fill in His own good time. "Here I have the power but not the wish to stay; there I wish to be but cannot; both ways, miserable."[17]

We might ask ourselves whether we should regret the historical happenstance that Augustine came along too soon in history to have received the kind of medical treatment recommended by Dr. Joseph Biederman in a presentation to the President's Council on Bioethics about the use of psychopharmacology in treating children.[18] Had

Augustine's symptoms—his inability to attend to what he found boring, his impulsive behavior, his restlessness—been diagnosed as Attention Deficit/Hyperactivity Disorder (ADHD) early enough, he might have been treated with Ritalin and spared years of anxiety and discontent, years of feeling that "myself to myself had become a place of misery, a place where I could not bear to be and from which I could not go."[19] He might have been spared the "horrendous inner sense of restlessness" that Dr. Biederman discerns even in many who are well past childhood. It is true, of course, that—compared to many of his peers—the young Augustine achieved a good bit; nevertheless, as Dr. Biederman noted, "'dysfunction' is a relative term." Despite his achievements relative to those of others, Augustine may not have been able to use his "intellectual ability to the fullest." Given that, in Dr. Biederman's words, "academic progress is a fundamental passport for a good life," perhaps with treatment Augustine could have lived out his life in Rome as rhetorician for the well-known and powerful, rather than in Hippo, carrying out the thankless duties of a bishop in the struggling African church. But then, of course, we would not have his *Confessions.* How easily we forget that one may become a philosopher through loneliness or discontent.

I exaggerate—but in search of an important point that the work of the President's Council on Bioethics may help bring into focus. In *Beyond Therapy: Biotechnology and the Pursuit of Happiness*, a report released in 2003, the Council noted that powerful stimulant drugs used, for example, to treat ADHD "have the capacity to enhance alertness and concentration in children without any symptoms whatsoever."[20] That is, the primary focus in *Beyond Therapy* is on "what it means *in general* [and not only in clear cases of over-treatment or under-treatment] to seek better or better-behaved children by pharmacological means."[21] A concern about medicalizing childhood is a concern about the means used to achieve an undeniably desirable end: happy, successful, and well-behaved children. And the use of drugs both to treat childhood disorders and to enhance children's accomplishments invites us to think more carefully about the meaning of childhood; for, we cannot know what we should want for our children unless we reflect upon what it means that they are children.

THE MEANING OF CHILDHOOD

There are places where *Beyond Therapy* may take an insufficiently nuanced view of childhood—as when, for example, it suggests that "the sweetness, freshness, and spontaneity of life are available in their purest form only to the as-yet-unburdened young."[22] Children have problems and anxieties aplenty, even when their problems fall short of any diagnosable psychiatric disorder. They are often very far from feeling unburdened, as Augustine's story demonstrates so acutely. They already experience the longing of the restless heart, well before they can articulate it. They need, therefore, nurture and direction. But having recognized this, we in turn need to think about how best to help them deal with their problems and anxieties as they mature.

If childhood is not as "unburdened" as the sentence from *Beyond Therapy* quoted above suggests, it is nonetheless true that we could easily impose burdens on children if we think of childhood only as a step on the way toward adulthood. So, for example, as part of his case for the importance of medicating children diagnosed with ADHD—indeed, even for denying entirely that our society has any problem of over-diagnosis or over-medication—Dr. Biederman asserted more than once to the Council his belief that "there is a connection between what you accomplish in school and where you end up in life" (or that "academic progress is a fundamental passport for a good life"). If a distinguished clinician approaches his young patients in that spirit, perhaps the Council was not wrong to worry that adult expectations might unduly burden children.

Gareth Matthews is that rare philosopher who has actually given thought to the meaning of childhood, and he has argued quite persuasively that we make a mistake if we think of it simply as a stage on the way toward adulthood.[23] Matthews characterizes developmental models of that sort—whether ancient, as in Aristotle, or more modern, as in Piaget—as "deficit" models. They describe childhood primarily in terms of those aspects of adulthood which childhood lacks (but toward which it is oriented): "The nature of a child is to be a *potential* adult."[24] Thinking that way, we are likely, with Dr. Biederman, to assume that children should naturally be anxious when taking a test (on which a bit of their future depends), rather than to wonder whether we should alter their school environment and our expectations in order to create less anxiety. We are likely to emphasize, with Dr. Biederman, how important it is that children be attentive even to material that is boring (lest, I guess, they end up with deficits, like Augustine's in Greek, which plague them the rest of their lives and make it difficult for them to live up to their potential). If we envision childhood as "an essentially *prospective* state, what is a good for a child can only be something that will contribute to its good in adulthood."[25] So if we find ourselves thinking in the terms of such a "deficit" model, the point of childhood will be simply to become an adult. And the point of adulthood will be—well, what? To be satisfied?

I do not want to press the matter farther than we should. Although I would resist—for reasons I will come to in a moment—describing the child as a kind of "incomplete" human being, there is surely some sense in which it is right to think of the child as immature and in need of nurture. Because we are not simply animals, parents must help their children to develop, as best they can, the full range of human capacities. Hence, the Council noted in *Beyond Therapy* that parents rightly want their children to become attentive, well-behaved, and well-adjusted. If they want this in part for reasons of "parental pride," they also want it in part because they think—with Dr. Biederman, reasonably enough—that "children who possess these qualities are more likely to succeed and flourish later in life."[26] We might even, as the Council notes, "find fault with parents who did not share" these aims, "at least to some considerable degree."[27]

It is very common to think of our lives in terms of a life cycle, for which the meaning of earlier stages lies partly in the way they prepare us for stages still to come. And thinking in terms of a life cycle might incline us, then, to picture the child as an incomplete human being—on the way toward full or complete humanity. But it is better, I think, to envision the child as (simply) "young" than as "incomplete." If we permit too much of the meaning of childhood to be gathered into its prospective and preparatory functions, we will see it relentlessly as a means to an end, and we are likely to try to take more control of its trajectory than is wise. Because we are not gods, parents are right to recognize limits to their work of shaping and forming their children.

Thinking in terms of the life cycle is not entirely mistaken, and it does point us toward a part of the truth about human life. It is not wrong to see childhood as, in part, oriented toward later stages of maturation still to come. That, of course, is why we are especially pained when someone dies in childhood; the full course of life's cycle has been cut short. Nevertheless, human beings do not "flourish" only in their adult state. A child can "flourish" precisely as a child; the meaning of his flourishing does not depend on future development or improvement. If the meaning of each stage of life were composed entirely of its prospective or preparatory functions, we would be hard pressed to see any point or worth to the last stages of life, when we have no further prospects.

For one image of life—that of the cycle—each stage of life has its meaning in terms of a trajectory of development that takes place over time. And this image does display a part of the truth about our lives. But equally true is the quite different image of a "circle" rather than a "cycle" of life. For that image every moment and stage of life, each point along the circle, is equidistant from its center.[28] It may be that the persuasiveness of this image—which begins to press beyond human dignity to what I have called personal dignity—relies upon some background religious beliefs. As all people, whatever their respective talents or accomplishments, are equal because equidistant from God, so each moment of life—whatever its stage of "development"—is equally meaningful and not just a step on the way toward some later stage. What, one might wonder, would heaven be like without children? The praise of God without the voices of children in the chorus? Each age, each moment of life, has its own distinct worth. Hence, we will miss certain truths about life if we think of our early years only as incomplete or as preparation for what is still to come.

In addition, when we think of childhood as only a prospective state, a means to the end of adulthood, we may try to take more control than is wise of the trajectory of a child's development. This is a constant temptation for both parents and teachers (and, dare we add, clinicians?). Paul Holmer taught for many years at Yale Divinity School. The story goes that at a banquet on the occasion of his retirement,

> after various faculty had spoken of him in glowing terms, the honoree was asked to respond. Holmer rose and read a short letter that he had received, saying that this was "my most treasured compliment to my teaching." The letter was from a young man who said that Holmer had changed his life, that his teaching had made a world of difference to his spiritual development.
>
> After reading the letter, Holmer sat down, saying only, "That young man is in jail serving a life sentence for murdering his father."[29]

Cutting still closer to the bone, John Ames, the 76-year-old preacher who is the narrator of Marilynne Robinson's *Gilead* and who is writing a long letter about life to his young son, recounts a sermon he had preached on Abraham and his sons Isaac and Ishmael. Abraham had been prepared to sacrifice Isaac at the Lord's command; Ishmael he sent off into the wilderness with his mother Hagar. Ames reflects that "any father, particularly an old father, must finally give his child up to the wilderness and trust to the providence of God. . . . Great faith is required to give the child up, trusting God to honor the parents' love for him by assuring that there will indeed be angels in that wilderness."[30] Perhaps, then, recognizing the limits of our ability to shape our children or to guarantee the course their development toward adulthood will take should lead us to discern our need not only for the image of the life cycle but also for the image of the life circle. Without it we cannot say all that needs to be said about the dignity of a child.

Thus, while recognizing that some children have serious psychological disorders and badly need medical treatment, we must also be alert to deeper philosophical questions—such as what it means to be a child—that will shape our thinking on diagnosis and treatment. If we think of childhood solely as a prospective state, and the child only as a potential adult or incomplete being, we are likely to see distracted, discontented, and difficult children simply as problems to be solved through powerful behavior-modifying drugs now available. But if we listen both carefully and critically to the medical experts and reflect upon those moments when they themselves are puzzled, we may become hesitant to medicalize too quickly such characteristics of children.

BEHAVIORS, CAUSES, EXPERTISE

In preparing *Beyond Therapy*, the President's Council heard from two physicians—Dr. Lawrence Diller and Dr. Steven Hyman—who discussed our attempts to improve the behavior of children through the use of psychotropic drugs.[31] At a later moment in its existence, studying more generally the problems of children in our society, the Council heard from Dr. Joseph Biederman and Dr. Fernette Eide and Dr. Brock Eide.[32] To consider some of the disagreements about both diagnosis and treatment that grow out of these discussions is, surely, to be invited to puzzle a bit over complicated questions about childhood.

Both Dr. Hyman and Dr. Biederman seemed relatively content with an approach to psychiatric diagnosis based less on an understanding of causes than on observation of (possibly symptomatic) behaviors. (Perhaps I should not describe Dr. Hyman as entirely "content" with this, for he emphasized that the science is for now "very early" and that observation of behavioral symptoms gives, at best, "reliability," not "validity.") While a disorder such as ADHD may have multiple causes—genetic, neurological, environmental—much about its origins remains uncertain. Nonetheless, so long as trained clinical observers, using standard behavioral criteria, agree in their diagnoses, we can—or so the hypothesis goes—regard their diagnosis as reliable, even if, in a sense, this tells us very little about what is actually happening to cause the behavior that concerns us. (I pass over here, though we should not ignore, the fact that an initial "diagnosis" often comes not from a trained clinician but from parents or teachers. If, as may often happen, a child then behaves more or less appropriately in the doctor's office, the doctor may rely on these reports of symptoms manifested elsewhere. Even Dr. Biederman, who is not inclined to grant the possibility that over-treatment is occurring, admits that the symptoms are variable, that they may occur in some contexts and not in others, and that a clinician may therefore simply rely on reports from others.)

Dr. Fernette Eide and Dr. Brock Eide were, by contrast, far from satisfied with this approach. In their view, many of the learning or behavioral problems that can result in a diagnosis of ADHD are caused by neurologically based disorders of cognition and learning (such as visual problems, partial hearing loss, dyslexia, language disorders) or by instruction that is not suited to the ways in which a given child may learn best or most efficiently. In their view, "many children who struggle in school do not have cognitive 'impairments' or 'abnormalities' in any absolute sense, but simply differences in learning-style—many of which actually render them well-suited for various adult occupations" (a viewpoint that is engagingly flexible when contrasted with the vision of academic progress as the passport to a good life). Yet, whatever the causes may be of such children's behavioral problems, the odds are increasingly good that they will be diagnosed with ADHD (and additional psychiatric disorders)—and, then, medicated to control their behavior. But medicating is not quite the same as "rearing," and in turning too quickly to a pharmacological fix we may treat the child as "just body," failing thereby to recognize the child's full dignity. Some of these children surely are genuinely disturbed; they need, and may benefit from, medication. But, at least in the clinical judgment of the Eides, it is less certain that others are really helped. They may be enabled to conform better to the schedules they are given, but it is not clear that they make measurable gains in learning.

Another factor complicating diagnosis came out especially clearly in the presentation to the Council made by Dr. Diller: namely, complexities buried in the concept of "impairment." Dr. Biederman, as I noted earlier, was concerned to argue that impairment and dysfunction are relative terms. One's achievements may be quite respectable and may even surpass those of many of one's peers (as, in my extended example, was true of Augustine) while, nonetheless, one is never really able, in Dr. Biederman's words, to use one's "intellectual ability to the fullest." Whereas Dr. Biederman seemed to have relatively little hesitation about the good of medicating such persons, Dr. Diller raised doubts and questions. He recounted, for example, an instance of medical students who had failed the national credentialing exam and had then sued the National Board of Medical Examiners—claiming that, because of an ADHD disorder, they needed unlimited time to take the exam. The court ruled against them, noting—reasonably enough it might seem to the lay observer—that, though they may be impaired in terms of their potential, as persons capable of completing medical school education they could hardly be said to be impaired when compared to the population in general. Any of us, even if our talents are many, may set the bar of our aspirations so high that we will almost inevitably experience anxiety; yet, it is hard to think that we are then best described as impaired. If such comparatively talented and capable people are afflicted with a disorder, it may have to do at least as much with their—or their parents'—desires and aspirations, their sense of what could count as a good life, than with any medical disorder—more, that is, to do with morality than with medicine.[33] Perhaps a clinician could medicate them and they might feel they had been "helped," but we should not be entirely indifferent to the distinction between help that works "with" them as they struggle to become persons of a certain sort and help that merely works "on" them, making adjustments of biochemistry.

These disputes are seldom, if ever, merely about facts. Deeper issues—which, for lack of a better term, we may simply call philosophical—are generally buried in the discussion, needing to be uncovered. Thus, for example, at least in my judgment, the Eides have a considerably richer and more nuanced understanding of the meaning of childhood than Dr. Biederman's rather monochromatic sense of childhood as an essentially prospective state, a stage on the way toward adulthood. No amount of medical expertise alone can tell us how best to think about childhood. Even a psychiatric clinician is not an expert on the meaning of the good life or how to live well.

In raising children, educating them, and shaping their behavior, we hope for certain results, and we look for means to achieve those results. But that is not all we should hope for. We should want something more than just a certain *result* (a disciplined, thoughtful, and cooperative child). If that were all we wanted, there would be little reason not to turn almost at once to medication. But our aim is not just to produce certain behavior, desirable as it, of course, is. We want a child who has begun to become a person of a certain sort—whose character is such that he or she is increasingly disposed to be disciplined, cooperative, kind. In search of such results, however, we cannot simply want to make a young Augustine less restless; for that discontented *eros* draws him, finally, toward God. Simply medicating to achieve a desired result would be to substitute "the language and methods of medicine for the language and methods of moral education."[34] What we want is an Augustine whose restless desire has been directed toward the goods of life in loves that are—to revert to the language he himself used—ordinate rather than inordinate.

From some perspectives, of course, such a position may seem out-of-date. Thus, for example, Dr. Biederman noted that in the past many would have described the case of a child who plays Nintendo for hours while seeming unable to do homework for two minutes as a "volitional" problem. But in Dr. Biederman's view, to characterize it as a problem of the will calling for moral discipline would be mistaken. Perhaps it sometimes is mistaken. But is that determination the province of any single group of experts? It raises deep questions about the nature of the self, the meaning of the self's involvement and agency in human activity, and the difference between achieving a certain result and becoming a person of a certain sort. As Council member Francis Fukuyama noted in the discussion that followed Dr. Diller's presentation, approaching these diagnostic issues too narrowly "reduces our understanding of moral agency" and "medicalizes a whole range of behaviors that traditionally were thought of as moral behavior," which had to be "socialized and taught." Of course, physicians must attend to such matters, but these are not questions for them alone to ponder.

In his presentation to and conversation with the President's Council, Dr. Steven Hyman leaned in the direction of a view like Dr. Biederman's, but the puzzling complexity of the phenomena eventually compelled him to draw back somewhat. There is, he said, no fundamental difference between using psychotropic drugs to affect the brain and using our disciplined experience to do so. Both are probably ways of accomplishing the same mechanistic aim: remodeling synapses within the brain. If that is really true, of course, it should make no difference to us which means we choose to reach the desired end when behavior needs to be altered.

Nevertheless, Dr. Hyman could not bring himself to be entirely indifferent, and he drew back somewhat from the implications of his own assertion. He preferred treatment approaches that do not immediately turn to drugs, fearing that doing so would send a message that this was the best way to deal with life's problems. The message to children would be something like: "Behavioral control comes from a bottle. We have the problem of anabolic steroids for the soul." If, however, there were really no significant difference between moral education and prescription drugs as means for reshaping behavior, it would be hard to know why Dr. Hyman should draw back from medicalizing behavioral problems (assuming the drugs being used are safe). Thus, his scientific/philosophical view of mind and brain stands in some tension with his reluctance to turn at once to medication as a solution to misbehavior and its accompanying discontents.

I do not say this with any intent to criticize him. On the contrary, honest puzzlement here is a sign of humane insight. It recognizes the peculiar character of human beings, those organisms whose mediated relation to the world (as Hans Jonas described it) gives rise to the capacity for reflection and the ability to act purposively. The action of such agents, unlike simple behavior, integrates body, mind, and spirit in ways that express the dignity of being human.

Because Ritalin functions as a kind of universal stimulant—enhancing alertness and attentiveness not only in those children with a true disability but also in all who take it—we are tempted to reduce a wide and complex range of behaviors (and behavior problems) to just one thing: chemistry of the brain, to be fixed with a pill. Surely, however, this is no better than obstinately insisting that the cure for all of a child's behavioral problems lies in disciplining the will. The philosopher John Wisdom told the story of a keeper at a Dublin zoo, who had a record of great success at breeding lions. When asked the secret of his success, he replied, "Understanding lions." Asked, then, what understanding lions entailed, he answered, "Every lion is different."[35] We should say no less of children—at least if we remind ourselves to think of them as human beings, those in-between creatures who can be reduced neither to will alone nor to brain chemistry alone.

CHAPTER FIVE

Loyalties

As living organisms, human beings exist only by remaining open to the world around them, a work which they carry out in ways much more complicated than do plants or, even, other animals. Because we are bodies, we have location and inhabit places; yet we are not "rooted" in place as plants are, and our capacity to perceive, desire, and reflect upon what is not immediately present to us means that we always to some extent transcend the place and community in which we are located.

It should be no surprise, then, that we experience an interplay—and tension—between the ties that bind us to certain people in particular and the more general duties we have toward all people. We would be inhuman—or angelic, which is, in its own way, inhuman—if we did not experience ourselves as especially obligated to those with whom either birth (nature) or breeding (a shared history over time) have closely connected us. Not to recognize those obligations would mean not to recognize ourselves as embodied, as located in a particular time and place. But it would also be inhuman—and demonstrate an inability to see that we are different from the other animals—if we did not recognize duties that we have to all members of the human family. The tension between the particular and the universal in our loves and our duties is grounded in the strange "in-betweenness" that characterizes human beings, who are neither beasts nor gods. And, of course, the longing for God that characterizes us means that we never belong to the whole extent of our being to any historical community. In a famous speech to "men of Athens," St. Paul said that God "made from one [ancestor] every nation of men to live on all the face of the earth, having determined allotted periods and the boundaries of their habitation."[1] Located in particular times and habitations, yet sharing a common humanity. It is no surprise, then, that the tension between particular loyalties and more universal loves and duties goes very deep into the structure of our religious tradition.

A CITY WHOSE BUILDER IS GOD

The history of ancient Israel is grounded in the call of Israel's God, an election that orients Israel to a land (and, eventually, to a city, Jerusalem) and to an identity distinct from other peoples. Against the backdrop in Genesis 11 of the story of the Tower of Babel, a failed project to create a universal community, Abraham is called at the outset of chapter 12. He is called to leave home and kin, but not in order to be a citizen of the world. He is to go to the land that God will show him, where he and his descendants will form a people. When his descendants are enslaved in Egypt, God once again calls them out to the land He has promised them. "When Israel was a child I loved him," the Lord says through the prophet Hosea, "and out of Egypt I called my son."[2] Among the purposes of the law of God, given to Israel through Moses, is that they should be a distinct people, set apart from the other peoples around them, and dedicated to the Lord.

Israel is given the promised land not simply as a place to occupy but as a place in which to share the way of life to which their covenant with the Lord has committed them. Indeed, the covenant is specifically understood as including and involving both "him who stands here with us this day before the LORD our God" (the generation that covenanted at Sinai) as well as "him who is not here with us this day" (generations still to come, who share a place, a history, and a way of life).[3]

This shared way of life brings with it special commitments to fellow Israelites. Thus, for example, the legislation recorded in Deuteronomy 23:19–20 with respect to lending at interest makes a clear distinction between insiders and outsiders: "You shall not lend upon interest to your brother, interest on money, interest on victuals, interest on anything that is lent for interest. To a foreigner you may lend upon interest, but to your brother you shall not lend upon interest."[4] Perhaps only such a clear distinction between insiders and outsiders can give membership in a community real power to shape identity; yet, of course, it also brings with it dangers. Both the power and the danger are unforgettably present in Psalm 137, a prayer of Israelite exiles in Babylon, far from the land where they had made their common life. The danger is obvious in the desire to see vengeance done to the Babylonians, but it is the longing for Jerusalem and the inability to sing the Lord's song in a foreign land that is most memorable. Any of us might be willing to risk the dangers in exchange for such an experience of belonging.

We sometimes suppose that the story of Israel in the Old Testament is essentially a tribal story, in which particular obligations to fellow Israelites are primary, and that it is in the New Testament that we find a faith which does not think of itself as tied to a particular place and which understands its mission in more universal terms. Although there is some truth in such a generalization, it is important to see that even—and also—in the Old Testament there are ways in which Israel's particular history points beyond itself.

The God of Israel is not a tribal god. Indeed, as Deuteronomy puts it, "to the LORD your God belong heaven and the heaven of heavens, the earth with all that is in it." True, this Lord "set his heart in love upon your fathers and chose . . . you above all peoples." Nevertheless, this one Who is "God of gods and Lord of lords" is "not partial." He loves not just Israelites who have covenanted with Him but also the "sojourner" in Israel. "Love the sojourner therefore; for you were sojourners in the land of Egypt."[5] Nor is it only the foreigner who temporarily resides within Israel's boundaries for whom this God has concern and with whom He has dealings. We may wish the prophet Amos had told us more, but he quite clearly suggests that the Lord, the God Who had delivered Israel from Egypt, has special relationships with other peoples as well.[6] Less obliquely, Moses reminds Israel in Deuteronomy that, just as the Lord had promised a land to Abraham and his descendants, so also that same Lord had given a land "as a possession" to the Edomites, descendants of Esau.[7]

Perhaps most important, the special bond of Israel's tribal brotherhood transcends (to some degree) "location" from the very start. The promise to Abraham in Genesis 12 is that the blessing upon his descendants will also be a blessing for all the families of the earth. The universality of the promise (at least in hope) is evident also in the fact that the servant who redeems Israel is a light to the nations, as Isaiah says:

> It is too light a thing that you should be my servant
> to raise up the tribes of Jacob
> and to restore the preserved of Israel;
> I will give you as a light to the nations,
> that my salvation may reach to the end of the earth.[8]

And, just as Amos had asserted that Israel's Lord had separate covenantal dealings with other peoples, even so the prophet Malachi is confident that "from the rising of the sun to its setting" the name of that Lord is "great among the nations."[9]

In the New Testament these hints of a loyalty that extends beyond particular and partial bonds become more pronounced. St. Paul characterizes the gospel he preaches as directed "to the Jew first and also to the Greek."[10] A central theme of the Gospel of Luke and its sequel, the Acts of the Apostles, is that "repentance and forgiveness of sins should be preached" in the name of the risen Christ "to all nations, beginning from Jerusalem." That mission may begin in Jerusalem, but it extends "to the end of the earth."[11]

Not only to the end of the earth but to anyone who may cross our path. The well known story of "the good Samaritan"[12] is powerful (and challenging) in part because the Samaritan's care for the beaten man has its ground not in any particular tie or loyalty but in their common humanity; indeed, it overcomes hostility rooted in their history. Likewise, in the Gospel of John, despite its pronounced emphasis upon love for fellow believers (fellow insiders), Jesus tells a Samaritan woman that a day will come—and, indeed, is now present in Jesus—when Jews and Samaritans will worship the Father not in their separate holy places but "in spirit and truth."[13]

Even as the story of Israel, while focusing on the particular bond shared by Israelites, pointed beyond itself to a more universal community, so also the story of Jesus and His followers, though directed to any and every neighbor and to the ends of the earth, is not a disembodied or purely spiritualized story.

The Word (*logos*) of God that is present in and upholds the entire creation is the Word made flesh in Jesus of Nazareth, a particular man in a particular time and place. Even when given new life after death, He does not leave the body behind. Only in Him, the prologue to the Gospel of John states, do we see the Father.[14] And when God ends history as we know it and creates a new heaven and a new earth, the City of God that comes down "out of heaven from God, prepared as a bride adorned for her husband" is called the "new Jerusalem."[15] The mission that began in a particular place, Jerusalem, only to spread to the ends of the earth, even when spiritualized as a new creation retains its embeddedness in time and history. It is still Jerusalem, even if a new, spiritual Jerusalem.

Granting that particularity has not been left behind, however, it is true that something decisive has happened. If a particular territory had been integral to Israel's understanding of its relation to the Lord, it now ceased forever to be a requirement of Christian faith. "Scripture tells us," St. Augustine writes, "that Cain founded a city, whereas Abel, as a pilgrim, did not found one. For the City of the saints is up above, although it produces citizens here below, and in their persons the City is on pilgrimage until the time of its kingdom comes."[16] Belonging to that city whose builder and maker is God, these people are something like the sojourners who lived in ancient Israel. They live there, they are located in that particular place, but they do not belong there to the whole extent of their being. Located there, they owe it their loyalty. Not belonging there to the whole extent of their being, their loyalty must always be less than ultimate. Any acceptable understanding of the ties and loyalties that characterize human life must strive to do justice to each of these truths.

THE "ACCIDENTS" OF TIME AND PLACE

As an example of failure to manage these intricate simultaneities, we may consider the following sentence from an essay by Martha Nussbaum: "The accident of where one is born is just that, an accident; any human being might have been born in any nation."[17] What this could actually mean, or whether it means anything precise, I am unsure. If "any human being" means "any particular human being," then her statement can be true only if time and place (the result of our embodied condition) enter not at all into a person's identity. It can be true only if our identities are entirely untouched by the fact that we are bodies. Only then could it be the case that having been born in Tehran rather than Indianapolis could make no real difference in who one is. Nonetheless, Nussbaum is quite insistent that to regard national boundaries as "morally salient" gives to what is "an accident of history a false air of moral weight."[18] To treat what is "morally arbitrary" as if it really mattered is to suppose that we could "magically" transform accidents of time and place into morally significant features of life.[19]

A charitable reader must assume that she does not really mean this. And, in fact, later in the essay, allowing that in order to be citizens of the world we do not have to relinquish "our special affections and identifications," she grants that "we may think of our identity as constituted partly by them."[20] How shall we hold together a cosmopolitan loyalty to whatever is good, which transcends all communal boundaries, with more particular allegiances and commitments? How, to take an obvious example, shall we justify the special care and concern we give our children?

Nussbaum's answer is a standard one, though no more adequate for being standard. Child care will be done poorly "if each thinks herself equally responsible for all." It is good for children on the whole that each of us is especially responsible for just a few of them. "To give one's own sphere special care is justifiable in universalist terms, and I think this is its most compelling justification."[21] This argument—which we might characterize as an attempt to "build down" from universal obligations to particular loyalties—does not persuade. At best it might suggest that particular adults should be given randomly assigned children to rear (rather than, say, rearing them all communally as wards of the state), but it cannot account for our sense that we should raise particular children who are ours, rather than just a few, generic and randomly assigned, children. It cannot, that is, account for our sense of the family bond—for the way both birth and breeding bind us in particular ways not governed by what makes for efficiency in child care.

There is, however, one angle of vision—the perspective of the Eternal—from which our particular location in time and space might be said to be "accidental." For all times and places are equidistant from God. From that perspective it might be hard to see why our loyalties and loves should be of any decisive moral significance. Taking this God's-eye view as best we can is instructive. It teaches us that the special allegiance we rightly feel to those bound to us by nature and history does not make them more worthy than others, nor can it justify any and every special preference to which we might be drawn. Unless there are limits to such preference we will have lost the peculiar duality of our in-between condition.

But, of course, from the perspective of the Eternal it is also true that there are no accidents. The accidents of time and place are built into our creaturely condition and should be understood as God's doing and gift—neither treated as ultimate, nor scorned as morally arbitrary. How to hold together within the moral life the truths that our special moral relations both are and are not "accidents" is, of course, no easy task.

BUILDING DOWN, AROUND, AND UP

The legitimacy of particular loyalties has occupied the attention of philosophers in recent years. A sense that the world we experience is a more global world, a related sense of cosmopolitanism that is leery of particular patriotic attachments, a liberal egalitarianism that draws back from anything suggesting special benefits or privileges—all of these combine to make some people skeptical of any claim that we are bound by special (and more stringent) duties to family, friends, near neighbors, fellow believers, or our country.[22] A more impersonal moral standpoint, less decisively governed by attachments to which the possessive pronoun "my" is naturally affixed, seems required. National patriotic sentiment, in particular, has seemed suspect and in need of special justification. And, to be sure, attachment to one's country is less "bodily" than attachment to one's family; it is grounded less in nature than in history. Nevertheless, certain kinds of historical ties—shared history, language, culture—enter deeply into our identities, and anyone not in the grip of a theory is likely to give them considerable moral weight.

How to justify that moral weight is not easy to say, however, especially if we grant that it is our nature not only to be located but also to be free (at least to some extent) of every particular location. It is our nature to be connected by special ties of birth and breeding with certain people; it is also our nature to be members of one human family. "God created man," Augustine writes, "as one individual; but that did not mean that he was to remain alone, bereft of human society. God's intention was that in this way the unity of human society and the bonds of human sympathy be more emphatically brought home to man, if men were bound together not merely by likeness in nature but also by the feeling of kinship."[23]

Samuel Scheffler has written of "a deep and persistent tension between these two features of our moral thought" and has suggested that we may not be able to achieve a "unified moral outlook."[24] Alasdair MacIntyre has argued that an impersonal moral standpoint and a morality which makes place for the "loyalty-exhibiting virtues" are "two rival and incompatible moralities."[25] The tension is, I agree, ineradicable in this life, for it is grounded in the two-sidedness of our humanity. Still, however persistent this tension may be, we can at least characterize four different ways of dealing with it.

One approach is less a way of dealing with the tension than of setting it aside. We could adopt what might be called a "Franciscan" attitude, finding in any and every person we encounter a human being whose very humanity calls forth our loyalty—an instance of that generic humanity toward which we have duties and which we are to love. Understanding our duties in this way, we could not make exceptions for those specially bound to us by nature or history; for they too are only instances of the humanity we are universally to love. Thus, Jesus' announcement of the kingdom of God and the allegiance it demands can take the form of seeming to question even the closest of our attachments: "Whoever does the will of God is my brother, and sister, and mother."[26]

Powerful as such a way of life may be, there is something disembodied—perhaps angelic or godlike, but still disembodied—about it. Of course, put in Christian terms, such a Franciscan spirit means to attempt to belong to a new body, a new community, that is not yet fully realized here and now. It aims at living here and now the life of the promised new creation, and in that sense it is the completion rather than the renunciation of the special ties we experience in history. "There is," Karl Barth writes, "an orphaned state required for the sake of the kingdom of heaven, in which a man who like all others is the child of his parents must symbolise with his being and action the present but hidden creation which is not a mere prolongation of the old, but the new creation in relation to which the old has already passed away."[27] But believers themselves have not thought that such a way of life is (here and now) for everyone.

There can also be philosophical versions of such an attitude, though we may find it strange to describe them thus. Unaccustomed as we are to thinking, for example, of John Rawls as Franciscan, a requirement that we try to think from within an original position—which is precisely a position without location or identity—is an attempt to overcome in thought the constraints of time and place as they shape our identity and attachments. This too has a disembodied quality, but without the hope that provides real energy. If Franciscan we must be, it might be better to fast for a time or endure a freezing vigil in the snow than to occupy the original position.

There remain three ways of recognizing the tension while trying to bring our particular and our universal duties into harmony. One way is to begin with the duties we have to all human beings (abstracted from any particular associations) and attempt to *build down* from them and make place within them for particular attachments and loyalties. From this angle we think of special ties simply as specifications of the bond we have with all but, given the limits of our finite condition, cannot enact equally with all. This is essentially the kind of move made by Martha Nussbaum. Surely there is something to it. Given the limits of human life, it is simply impossible for us to live equally for all others. Life is likely to flow more smoothly, and, in fact, the good of more people is likely to be served, if each of us gives special attention to some others who are closely connected to us by either natural or historical ties.

Attractive as this approach may seem—making primary our equal humanity and justifying special preference only on the basis of that primary commitment—I doubt that it can be made convincing. We noted earlier how inadequate was Nussbaum's justification of special preference for one's children: the argument that children will be raised better if each of us is responsible for a few of them. An attempt to build down from world citizenship to such particular responsibilities, grounded in the limits of our finite condition, can give no reason why my special care should be devoted to children bound to me by natural ties rather than random assignment. The very argument transforms the nature of the particular bond and, hence, does not offer justification for what we actually mean by such bonds. No one loves his children for this reason—as an instance of children taken generally. In order to be responsible agents in the world—at least, responsible human agents—we must actually be relieved of the kind of overall responsibility appropriate only to a god. For us, "being responsible for absolutely everything undermines the very idea of responsibility."[28]

The same is true, if less obviously so, of patriotic attachment to one's country. This becomes especially clear if we focus on what has always been a problem for liberal theory in politics: willingness to die for one's country. If citizen soldiers are willing to risk such sacrifice only when they are first convinced that their country's cause is just, "measured by some standard that is neutral and impartial relative to the interests of their own community and the interests of other communities," the kind of allegiance any community needs to survive will be lacking.[29] The existence of citizens of the world, if there really are such, is parasitic upon the presence of others whose loyalties are more particular and personal. This does not mean that a disinterested respect for the legitimate concerns of citizens of other countries is not also important. Were we to ignore those claims upon us we would, of course, also be denying an important aspect of our humanity. But the binding force of shared nationality itself needs no more universal ground. "I may without self-righteousness or hypocrisy think it just to defend my house by force against a burglar; but if I start pretending that I blacked his eye purely on moral grounds—wholly indifferent to the fact that the house in question was mine—I become insufferable."[30]

If building down from universal and impersonal duties to special bonds that obligate us more stringently does not seem to work, common sense is likely to suggest a different approach—what can be called *building around*. Loyalty to those bound to us in particular ways is fitting and, indeed, to be expected and approved as integral to a flourishing human life. But, of course, such preference may sometimes exclude or ignore others in ways that are wrong or unfair. Because our personal loyalties can easily go awry in this way, our more universal duties must build a protective hedge around them, setting limits on what may be done to outsiders for the sake of those to whom we are bound in special ways.

This approach does, as I noted, have common sense on its side, and, in fact, it may be the most sensible approach to take if we bracket religious faith entirely from our deliberations. Samuel Scheffler, a philosopher whose reflections on loyalty have been sustained, rigorous, and clarifying, seems finally to opt for something like a "build around" model. He recognizes that there is a deep tension—perhaps a permanent tension—in ordinary human experience between our acknowledgment of "associative duties" and our belief in a "principle of equality." We have associative duties to others with whom we have close personal relationships or who are members of social groups significant for our own identity. The source of these duties "lies neither in our own choices nor in the needs of others, but rather in the complex and constantly evolving constellation of social and historical relations into which we enter the moment we are born." Yet, at the same time, the principle of equality "asserts that all people, however varied their relations to us may happen to be, are nevertheless of equal value and importance."[31]

How shall we reconcile and unify these—or, if that is too strong, hold them together within an individual's moral life? Scheffler's suggestion is that our "special responsibilities need to be set within the context of our overall moral outlook and constrained in suitable ways by other pertinent values."[32] This formulation seems to recommend that we allow certain general responsibilities (that grow out of the principle of equality) to set limits to what we understand to be required or permitted by our associative duties. It does not attempt (as the "building down" approach does) to derive particular associative duties from any more general principle, as if they were simply "instances" of some more general obligation. Those special loyalties have their own moral urgency that is not derived from the more general responsibility we bear toward all people. That urgency must be honored even while it is sometimes restrained and limited.

Scheffler doubts that any more unified moral outlook is available to us, and perhaps that is true. Still, it is useful to recognize the sense in which each aspect—the particular and the universal—might be said to come first for this approach. From one angle it makes sense to say that certain obligations we have to all people (not to injure them deliberately, to respect their personal freedom and bodily integrity) come first and set limits on all that we do. From another angle one might note that only because we feel loyalty to those attached to us in particular ways do we understand why others might have similar attachments that deserve our respect. To be human is to live within this tension and to work out the problems it presents as best we can.

If, however, we are reluctant to settle for affirmations of tension and are in search of a more unified moral outlook, we might attempt to *build up* from particular loyalties to an allegiance to every human being. I suspect, for a reason I will come to in a moment, that it is religious thinkers who are most likely to be drawn in this direction—a direction that makes the virtue of hope central to the problem of conflicting loyalties. Neither attempting to derive special bonds from more universal commitments, nor accepting that our allegiance to all human beings is little more than a negative boundary-setting principle, this last approach begins with particular loyalties and sees in them a training ground, a place where we may gradually learn how to honor and respect the worth of any human being. Learning to love a few, we may learn to love more generally. Moved by particular loyalties to detect worth even when it is not always in evidence, we may become able to see it also in those to whom we have no special attachment.

Suppose, for example, that we speak—as I have here and there in this chapter—of allegiance or obligations to all members of "the human family." Or suppose we speak of "friendship to all the world." Or suppose we speak of "citizens of the world" or "a global community." How can we know what such locutions mean? And from where do they acquire a certain intensity? A notion of "the human family" can have meaning for us only because we begin with an understanding of particular families and our attachments within them. "Friendship to all the world" could mean nothing to us did we not experience particular friendships. And, likewise, we can think of ourselves as potential citizens or members of a universal society only because we have known what it means to belong to smaller political communities and to have fellow citizens. Our loyalties begin closer to home, and we then extend their meaning in ever widening circles.

Jesus' statement that "whoever does the will of God is my brother, and sister, and mother" is among His "hard sayings," seeming to demand that we renounce loyalties that are very dear to us in the name of some more all-encompassing commitment. Yet, the saying is hard only to those who have begun with particular loyalties such as the family bond. Only they can actually be called to renounce a closer personal bond for the sake of a larger scope for love and responsibility. Thus (echoing I John 4:20) C. S. Lewis writes that "those who do not love the fellow-villagers or fellow-townsmen whom they *have* seen are not likely to have got very far towards loving 'Man' whom they have not."[33]

There is, however, more to the *building up* view than the claim that we must begin with particular loyalties if we are to make sense of our talk about more extensive allegiances. There is also—or, at least, there may be—a certain dynamism to this approach. The special moral relations that mark our identity—and should mark the identity of embodied creatures such as we are—need not, taken by themselves, be our final destination. For they are also a kind of training ground in which we learn the joys and the responsibilities of attachment to some and become more able to draw still others within the scope of our loyalty. To speak of particular bonds of loyalty as training grounds is to interpret them teleologically, as oriented toward some further purpose—and we are most likely to think of them in this way when we discern a providential hand at work in the way they shape and form us. Thus Augustine writes that the servants of God "have no reason to regret even this life of time, for in it they are schooled for eternity."[34]

We are constantly tempted to make of our particular attachments —family, community, country—both too little and too much. Too much—when we see in them not only the beginning but also the termination of our duties and our capacity for loyalty. Too little—when we suppose that we can be citizens of the world first, belonging nowhere in particular, without having learned somewhere what it means to be loyal. What we need—and what the approach of *building up* most successfully offers—is a language that can affirm the moral importance and necessity of particular loyalties, embedded in our nature and history, without depriving those attachments of that still greater *telos* best described not as cosmopolitanism, nor even as the human family, but as children of the one God. In that way we affirm and live within the two-sidedness that characterizes our humanity.

CHAPTER SIX

Death

HUMAN BEINGS ARE NOT immortal gods, and the trajectory of our lives moves inescapably toward death. But what if we were offered an indefinite continuance of this life—a kind of earthly immortality? That is the choice offered the hero Odysseus in the *Odyssey*, as he struggles to return home after the Trojan War. As the poem begins, we are told that he is a captive on the island of Ogygia, kept there by the nymph Calypso, who has taken him as a lover and offered him an immortal life with her.

> By now,
> all the survivors, all who avoided headlong death
> were safe at home, escaped the wars and waves.
> But one man alone . . .
> his heart set on his wife and his return—Calypso,
> the bewitching nymph, the lustrous goddess, held him back,
> deep in her arching caverns, craving him for a husband.[1]

But it is Odysseus' desire and destiny to return to his home and his wife Penelope, and eventually Zeus orders Calypso to release him. Even as she does so, however, sending him back to Penelope, Calypso cannot help noting to Odysseus that "I just might claim to be nothing less than she, / neither in face nor figure."[2]

> "Ah great goddess,"
> worldly Odysseus answered, "don't be angry with me,
> please. All that you say is true, how well I know.
> Look at my wise Penelope. She falls far short of you,
> your beauty, stature. She is mortal after all
> and you, you never age or die . . .
> Nevertheless I long—I pine, all my days—
> to travel home and see the dawn of my return."[3]

Odysseus is a man, not a god, and to live forever with Calypso—however pleasurable and desirable—would be to deny the dignity of his humanity, a dignity fully on display in this scene.

If human beings are not gods, neither are they beasts, moved simply by a desire to survive and enjoy more of the life we know. When Ken Burns produced his much-acclaimed series on the Civil War, one of the most powerful moments for many listeners was the reading of a letter written by Major Sullivan Ballou of the Second Rhode Island regiment of the Union Army to his wife, Sarah. Knowing that he was soon to go into battle and might not return to her or to his sons, he wrote to Sarah: "I have sought most closely and diligently, and often in my breast, for a wrong motive in thus hazarding the happiness of those I loved and could not find one. A pure love of my Country and the principles I have often advocated before the people, and 'the name of honor that I love more than I fear death' have called upon me, and I have obeyed."[4]

Ballou was soon to die; yet, here we see him flourishing as a human being—one whose dignity is displayed in how, rather than how long, he lives. Odysseus struggles to go home to his wife rather than live forever with Calypso, for he knows that human beings are mortals not meant to live endlessly the sort of life we now experience. Ballou willingly risks never returning to his wife, yet for reasons not unlike Odysseus's. He knows that there is something more compelling than simply surviving to enjoy more of this life (even in its undeniable sweetness).

DEATH AS FRIEND

"It's not immoral to want to be immortal." That was the title given to an MSNBC commentary by bioethicist Arthur Caplan, who was entering into an ongoing discussion about the wisdom of research intended to delay aging and extend the lifespan.[5] The issue—which has to do with extending not the average but the maximum lifespan—is, of course, too complex to be dealt with in a slogan. Caplan himself notes that there is no foreseeable future of ours in which immortality would be a genuine possibility. Hence, he says, "the debate is about living a lot longer than we now do, not living forever." True though that almost surely is, it misses another truth: that human desires know few limits. Trying to decide whether living "a lot longer" would be good for us is a fruitless undertaking unless our reflections are grounded in some deeper understanding of what sort of creature the human being is.

As our way into such reflection, we can examine an essay that itself goes much deeper than slogans: Leon Kass's "*L'Chaim* and its Limits: Why Not Immortality?"[6] The "core question," Kass articulates as follows: "Is it really true that longer life for individuals is an unqualified good?"[7] I'm not sure this is the best way to put the central concern. "Is it really true that dying is an unqualified evil for us?" comes closer to the way I would want to put it. Perhaps there is little difference between the two formulations, though I will try to suggest that there is.

Movement toward death is built into organic life and, apart from divine intervention, will inevitably occur. Even if our bodies do not break down—if, that is, no disease overcomes us—they will, sooner or later, wear down, losing their capacity to carry on those exchanges with the surrounding world that are the secret of life. They do, of course, often break down before they wear down. Disease, disaster, and evildoing add misery to inevitability and sometimes make death seem welcome. But even apart from such miseries, organisms die, and we may therefore ask what we should make of that fact.

Kass tends to think of it as a blessing, and he offers a range of reasons why one might think that. All of them turn on some way in which the limit of death is needed for life to be sweet and meaningful. Thus, he wonders whether life would be serious, would really seem to matter, if we knew it had no limits—if we did not need, as the psalmist says, to learn "to number our days."[8] Or again, our appreciation of much that is beautiful may depend on a sense that it is transitory and impermanent. Still more, virtue requires at least a readiness to sacrifice even one's life in pursuit and defense of what is good, and this would be impossible were we invulnerable.[9]

There is something to each of these reasons, though none of them fully persuades me. To see why, we can think through a fourth reason Kass offers to see death as a blessing: Our ability to remain interested and engaged in life depends upon our knowledge that it will end. Could we, for example, sustain indefinitely our interest in sports, in children, in vocational achievements? Of course, the miseries of life brought upon us by disease, disaster, and evildoing are likely to mar our enjoyment of life's pleasures and activities, but, if we bracket that sad fact, it is less clear that there is reason to become satiated with the good things of life. To be sure, as we wear down, the range of goods we are capable of enjoying becomes more restricted, but that is not the same as losing the will to take an interest where we can. "I once heard," Daniel Callahan writes, "someone's elderly grandfather described as a man of great energy and activity who, as he aged, had to live, because of illness and aging, within a smaller and smaller physical radius. Yet, even as that radius narrowed, first to the yard he could not leave, then to the house he could not leave, then to the room he could not leave, and finally to the bed he could not leave, he adapted to each smaller world, making of it with good cheer whatever was possible."[10] That is an example of human dignity from which we all might learn.

Moreover, the idea that life could at some point lose its ability to engage and interest us suggests that the creation might fail us. In one sense, of course, this is surely true; for the desire that moves us is not, ultimately, for more of this same life. It is a desire for God, and every good we enjoy is an intimation of a Goodness more lasting than any we now experience. But in that sense the creation fails us—fails to satisfy the heart's deepest longing—not only when we have lived to a ripe old age but, also, at every moment along the way.

To be satiated with life suggests a failure not in the creation but in us. Of course, activities in which we engage or purposes at which we aim solely as means to other goals may lose their capacity to interest us if the goal is attained (or becomes clearly unattainable), but this is not true of other goods, sought for their own sake (such as work, play, aesthetic experience, friendship, worship). Perhaps there are some activities, which, though innocent, do not merit our indefinite pursuit or enjoyment, though I would be hesitant to assert this with much confidence. As Screwtape advises Wormwood: "Never forget that when we are dealing with any pleasure in its healthy and normal and satisfying form, we are, in a sense, on the Enemy's ground. . . . [I]t is His invention, not ours."[11]

In general, I am inclined to think there is no reason to suppose that our sense of engagement in the good things of life—or our capacity for seriousness, our appreciation of beauty, our commitment to virtue—depend on the transitory or fragile character of this life. Indeed, were they so dependent, the very notion of heaven would be incoherent. All genuine goods have a kind of open horizon; there never comes a point at which we cannot enter into them yet more deeply.[12] C. S. Lewis captures this beautifully at the end of *The Last Battle.* When all the faithful Narnians and the children from our world are drawn into Aslan's world, the refrain becomes: "further up and further in." It strikes them that Aslan's world is much larger than it had seemed when they were still outside it, and Lucy puzzles over this with Tumnus the Faun.

> "I see," she said at last, thoughtfully. "I see now. This garden is like the stable. It is far bigger inside than it was outside."
>
> "Of course, Daughter of Eve," said the Faun. "The further up and the further in you go, the bigger everything gets. The inside is larger than the outside."[13]

The capacity of any genuine good to draw us and more deeply engage us knows no limit. If we become satiated, the failure is ours.

If—apart from the misery brought upon us by disease, disaster, and vice, which misery we have bracketed in our discussion—there is no reason to grow weary of the good things of this life, we might reverse the answer Kass gives to his "core question": "Is it really true that longer life for individuals is an unqualified good?" Yes, it is. But there remains my reformulation: "Is it really true that dying is an unqualified evil for us?"

To that question our first answer must be "no" (though a "no" that will eventually require a second, qualifying, answer). As the hart pants for cooling streams, so also, the psalmist writes, our souls thirst to behold the face of the living God.[14] If the deepest desire of our hearts, intimated by all our loves, is for God, the fullness of Whose presence cannot be enjoyed in this life, then more of this life—even its indefinite prolongation—could not possibly satisfy that desire. Thus St. Augustine, in the *Confessions* characterizes the inability of this life to fulfill us. "Here I have the power but not the wish to stay; there I wish to be but cannot; both ways, miserable."[15] Even as organic life is taken up into a higher, personal, life in human beings, so human life can become something qualitatively higher only as it is taken up into the divine life—a mysterious fulfillment that can come only the other side of death. If we characterize human dignity only in terms of nature and not also in terms of destiny, we shut our eyes to the sense in which death is not an unqualified evil for us.

If this is true, then it may also be true that aging is good for us. We resist such a conclusion, of course, and would like to live the whole of life without any diminishment of our capacities—and then simply to die suddenly, as if falling off a cliff, while still at the peak of our powers. But if we never experienced any lessening of our powers, would we be willing to accept the truth that more of this life is not what we really seek? Almost surely, many of us would not. It is aging that keeps us from imagining that everything our hearts desire could be given through more of the same kind of life. And it is aging, wearing down, that enables us to cultivate within ourselves the capacity for self-giving and self-sacrifice that makes place for those who come after us. To grow old, to wear down, even to die—and to know and acknowledge this as part of life's trajectory—is fitting for a creature who is neither beast nor god, and whose dignity consists in being human.

If dying is not simply an evil for us as individuals, is it, then, a good that we should at some point—perhaps at any point that suits us—embrace? That is a different question, and it requires a different answer.

DEATH AS ENEMY

Among those things that Filostrato saw as "most offend[ing] the dignity of man" was death. And, in fact, an attempt to sustain the "life" of a kind of disembodied being lies at the heart of the plot in *That Hideous Strength.* For good reason, then, even while piling up considerations that might move us to think of death (when timely and not premature) as a blessing, not only for the human species but also for its individual members, Leon Kass was quick to add that he was not suggesting "there is virtue in the particular *event* of death for anyone or that separation through death is anything but painful for the survivors."[16] It seems right that he should gloss his argument in this way. A wife who has lost her husband mourns. One of the deepest human attachments within her has been dealt a cruel blow. She can remarry, but can she simply "replace" her husband with another and in that way repair the loss? Hardly. To see why not will take us across the border that distinguishes thinking about the dignity of being human from thinking about the dignity of the person.

Although from one angle of vision we can see how death is not simply an evil for us—how it keeps us from imagining that we could find fulfillment in an indefinite continuation of the years of this life—when we shift that angle to focus not on the way we share in being human but on the uniqueness of each person, death suddenly appears as an evil and an offense. The husband lost in death is not only an individual in Kierkegaard's sense of a "distinguished person"—one with capacities and attachments that could perhaps be replicated. He is also Kierkegaard's "single individual"—known by name to God and marked as unique by that relation. He cannot be replaced—not partially, not at all. For he lives not only within the orbit of his earthly attachments but also in relation to God.

Each of us is singled out equally as one whom God knows by name. Death puts an end—so far as we can discern—to a unique and irreplaceable person. Unsurprisingly, then, we sense its offense and are driven to try to master it—to find a way to make the transition from death to life without actually passing through it. What is understandable is nevertheless also illusory. We cannot master the very event that announces to us our lack of mastery.

Still, it is not bad that we should struggle against death as an enemy. However natural and inevitable a part of organic life death may be, we should never think of it simply as a good to be embraced, whether for ourselves or for another, since the person who dies is equally but uniquely one of us. We should, therefore, resolve to reject the idea—even when packaged under the rubric of "death with dignity"—that it could be right deliberately to aim at our own death or that of another person, however difficult it may be to sustain this resolve when the hardships of life are great.

I have argued above that the loss of capacity we experience as we grow older and wear down is no violation of our human dignity—and, in fact, is integral to the dignity of being human. It reminds us that fulfillment is to be found elsewhere. But the misery we sometimes endure when attacked by disease is a different matter. One may perhaps bear it with a kind of dignity, but the misery itself undermines our hope to flourish as human beings. We are tempted then to conclude that continued life, now so burdensome, no longer has any value for us—or, at least, that the game is no longer worth the candle.

That temptation should be resisted. Even if, as might at some point be the case, my life no longer seems to have any value *for* me, it will still have value (we might say) *in* me.[17] That is, my dignity as a person —which is quite a different thing from the "worth" of my life to me or to others—still demands respect, not comparative assessment. Were my person just a "vehicle" for the realization of different valued states of affairs in the world, then the person might well be cast off when sober calculation suggests that the balance of good and ill realized in this life is no longer a favorable one. But that is not what it means to be a person.

After all, it would violate our dignity as persons if we were (freely and willingly) to sell ourselves into slavery, even if the payoff from doing so was considerable. In terms of value, we might profit from the transaction. But we would have done wrong to the dignity of the human person that inheres in each of us. Likewise, to decide that life is no longer worth living is simply to make a calculation about harms and benefits for me—quite a different thing from respecting the dignity of the person in me.

Although we may sometimes rightly give our life for the sake of another person, we cannot give ultimate authority over that life—cannot hand over our person—to another. Nor can we accept such authority over another's life, as if we were not equals. There are circumstances—in punishment, in war, in police work—that sometimes justify deliberately taking another's life, but a society can safely permit even these necessary activities only if it acknowledges that we do not, in the end, issue the final verdict on the "worth" of any person's life.

This does not mean that life may never be risked, as if we must all be librarians rather than police officers. Nor does it mean that respect for personal dignity requires us to do whatever we can to stay alive for as long as possible. It means only that we should not simply embrace the death of any person as a goal at which we aim. When we withhold or withdraw medical treatment for good reason—because it is no longer helpful, or because the treatment itself is very painful and burdensome—we do not choose death. We choose life. From among the life choices open to us, we choose one life, even if it is shorter than some that might still be available to us. When we engage in a risky or dangerous activity, we normally do so, not to aim at death, but in order to choose a life marked by certain goods. Respecting the dignity of the person makes central how we live, not how long.

We should not confuse this respect with kindness. A kindness that, moved no doubt by charitable concern, can think only in terms of harms and values, loses the still more fundamental language of respect and fails thereby to do justice to the dignity of the person. It is death as offense, as enemy, that forces us to realize that each of us is not just a member of the species but "that single individual"—one of us and equal to us—whom we must honor and respect. We are driven, then, to press beyond human dignity to personal dignity.

Personal Dignity

CHAPTER SEVEN

Confusions

IN MARCH OF 2008 the President's Council on Bioethics released a large volume titled *Human Dignity and Bioethics*. It is in some ways a strange product, consisting of essays by both Council members themselves and others who were invited to write for the volume, and displaying a wide range of understandings of the concept of human dignity. Its strangeness lies partly in the fact that the project grew out of an attempt to gain some clarity but ended in a cacophony of different voices.

In parts of its previous work the Council had made use of the language of human dignity, though, as some critics had noted, without fully analyzing or clarifying that language.[1] Thus, arguing that "dignity is a useless concept," Ruth Macklin criticized the Council's failure to provide an analysis of the concept of dignity it used.[2] James Childress, though without condemning dignity as a useless concept, agreed that the Council had "tended to invoke rather than really use the idea of human dignity" and had left it largely "unanalyzed."[3] One might, then, take the essays in *Human Dignity and Bioethics*, with their vastly different perspectives, as a confession of inability to bring clarity to the concept of dignity. Yet, even out of such different perspectives, an attentive reader who genuinely seeks clearer understanding and insight may sometimes find it.

As we turn from the explorations in the first part of this book, which explore aspects of *human* dignity, to the somewhat different language of *personal* dignity, it may be useful to examine an attempt to clarify which, in the end, only confuses. In the May 28, 2008 issue of *The New Republic*, Harvard evolutionary psychologist Steven Pinker published a long review of *Human Dignity and Bioethics* under the title "The Stupidity of Dignity."[4] Because Pinker is a relatively well-known scholar, who has written about human nature, we may gain some clarity through a brief exploration of his own lack of it. Where his misunderstandings do not turn on animus or willful misreading, they invite us to think about the distinction at the heart of my discussion—between the "dignity" that characterizes that in-between human creature who is neither beast nor god, and "dignity" as the standing of every human person, which calls for our equal respect. By analyzing just a few passages from his review we can illustrate this.

Pinker begins by noting some of the bioethical matters the Council has taken up: drugs to enhance cognition, genetic manipulation, anti-aging research, attempts to derive embryonic stem cells, and cloning-for-biomedical-research. And he comments: "Advances like these, if translated into freely undertaken treatments, could make millions of people better off and no one worse off. So what's not to like? The advances do not raise the traditional concerns of bioethics, which focuses on potential harm and coercion of patients or research subjects."[5] I pass by here the obvious fact that his characterization of the "traditional concerns of bioethics" is much narrower than any historian of the development of bioethics would recognize as accurate.[6]

The more important point is that thinking seriously about human dignity should compel us to ask whether freedom and consent are really all we ought to care about—to ask whether being human means nothing more than the freedom to shape and reshape ourselves, or whether it also means honoring the embodied character of our life and affirming some of its limits. Appeals to human dignity often assert that people may be wronged (when the human dignity they share is not respected) even if they are not in obvious ways harmed. And, in addition, such appeals assert that it is possible for us freely to consent to what demeans our shared humanity. These may be mistaken claims, but they are not obviously "stupid," and our willingness to take them seriously says a good bit about how we understand human life.

Thus, we need far more argument than Pinker provides if we are to accept his assertion that the concept of dignity is "just another application of the principle of autonomy."[7] This assumes that freedom (to shape and reshape our life) is the sole truth about human beings, and that, of course, reduces the complexity of our humanity, ignoring the significance of the body. Take a case (to which I myself had referred in my essay in *Human Dignity and Bioethics*) of the "sport" of dwarf-throwing. Why might we prohibit such sport, even if the dwarves consent and gladly participate? The only reason Pinker can imagine is that permitting the sport might encourage us to mistreat all dwarves (though he would like some empirical support offered to undergird such a claim). But we might have recourse to the language of human dignity precisely to suggest that the dwarves could be wronged (the human dignity which they share demeaned) even if neither they nor the larger society were in obvious ways harmed by such sport.

Pinker describes as "superb" an introductory essay by Daniel Davis in *Human Dignity and Bioethics*.[8] He should have read it with greater care. For one of the points Davis makes in his historical reflections on the development of bioethics is that the central principle of "respect for persons," though it includes respect for autonomy, cannot be reduced to that alone. And, in fact, the "persons" to be respected have often been understood to be all human beings—many of whom may not be autonomous at all, but all of whom have a claim on us for our respect.[9] The clash between these several ways of thinking about respect for persons is neither new nor ended, and reducing all moral "wrong" to violations of freedom or infliction (without consent) of tangible harms is to presume a conclusion rather than argue for one.

Continuing his depiction of the Council's work, Pinker comes directly to the concept of dignity. There are people who "demonize technology," people who worry about techniques for enhancing our capacities, and people who are repelled by certain manipulations of our biology.[10] "The President's Council has become a forum for the airing of this disquiet, and the concept of 'dignity' a rubric for expounding on it. This collection of essays is the culmination of a long effort by the Council to place dignity at the center of bioethics."[11] One feels at this point a certain need to introduce Pinker to the history of the second half of the twentieth century.

In 1948 the General Assembly of the United Nations adopted the Universal Declaration of Human Rights. Its Preamble begins with "recognition of the inherent dignity . . . of all members of the human family." The staying power of this language, and its application in particular to bioethical concerns, is evident in the Universal Declaration on the Human Genome and Human Rights, adopted by UNESCO in 1997, which likewise begins with an affirmation of the inherent dignity of all "members of the human family." It has even been suggested that "dignity" is "'the shaping principle' of international bioethics."[12] Such developments are well beyond the power of the President's Council to effect.

And whereas Pinker depicts the Council's appeals to dignity as springing from "a movement to impose a radical political agenda, fed by fervent religious impulses, onto American biomedicine," the direction of influence is probably quite the opposite.[13] Thus, for example, Pope John XXIII's *Pacem in Terris*, one of the greatly influential social encyclicals in recent Roman Catholic history, is utterly suffused with the idea that human rights are grounded in human dignity, an approach almost surely influenced by the rise to prominence of this language first in the political realm.[14] Indeed, *Pacem in Terris* characterizes the 1948 adoption of the Universal Declaration as "an act of the highest importance. . . . For in it, in most solemn form, the dignity of a person is acknowledged to all human beings."[15] Thus, although the President's Council might be flattered to think of itself as having so great an influence, the truth is that its recourse to the language of dignity is hardly new, hardly radical, and hardly fed by fervent religious impulses.[16]

This, of course, is only to get the history right (though to treat it as carelessly as Pinker does is no minor flaw). But this does not in itself demonstrate that dignity language is helpful or clarifying—or that it is not, to use Pinker's term, a "squishy" notion.[17] Some of that squishiness might have been overcome, however, had he been willing to enter a bit more sympathetically into a few of the essays in the volume he was reviewing. "We read [in these essays]," he writes, "that slavery and degradation are morally wrong because they take someone's dignity away. But we also read that nothing you can do to a person, including enslaving or degrading him, can take his dignity away. We read that dignity reflects excellence, striving, and conscience, so that only some people achieve it by dint of effort and character. We also read that everyone, no matter how lazy, evil, or mentally impaired, has dignity in full measure."[18] But of course. These are exactly some of the normative viewpoints that separate the essayists and that invite not dismissal of dignity as stupidity but careful reflection about it.

How can it be that slavery violates the dignity of the person enslaved but that nothing we do can deprive another person of his dignity? Perhaps these assertions are simply incompatible, but perhaps slavery demeans *human* dignity embodied in the slave, yet is utterly incapable of actually depriving him of the *personal* dignity he shares equally with all of us. The language of dignity serves as a placeholder for a vision of what it means to be human, a humanity that is subverted if we think of human beings simply as animated commodities capable of being owned by others. And the language of dignity also affirms—in a manner to be developed more fully in the next chapter—that we live a lie if we suppose that among two people made equally for life with God one can be master of another. Thus, one might object to slavery both on the ground that it violates *human* dignity and on the ground that it fails to recognize the dignity of the person enslaved.

How can it be that we use the language of dignity to point to certain activities attainable by only some of us while also using it to assert that even the weakest and least capable of us is equal in dignity to the talented and able? Perhaps these claims are simply incompatible, but I think not. Those whose behavior and accomplishments are characteristic of humanity at its best give to all of us intimations of the dignity of human life. But, at the same time, even one who thus flourishes can only share equally the dignity that belongs to all persons, whatever their capacities, achievements, or character.

In general, we cannot make sense or achieve sympathetic understanding of appeals to the concept of dignity unless (whatever terminology we choose to use) we distinguish human dignity from personal dignity, for they involve different sorts of claims. Nor can we understand the kind of claim appeals to dignity often involve unless we are at least willing to consider that respect for persons may mean more than refraining from harming them or failing to acknowledge their autonomy. We can illustrate this by thinking briefly about one of Pinker's examples with which most readers (including, I have to say, myself) are likely to find themselves in some agreement.

He writes that Leon Kass, who chaired the President's Council in its early years, had a "fixation" on the concept of dignity which sometimes carried him "right off the deep end."[19] As evidence for this he cites a passage in which, discussing eating, Kass writes that eating in public —and especially a "catlike activity" such as licking an ice-cream cone while walking down the street—falls short of truly human eating.[20]

On the matter of public licking of ice-cream cones, my own sensibilities are closer to Pinker's than Kass's. But from Pinker's passing comments one would never guess that Kass's discussion of licking ice-cream cones is part of an extensive examination of eating as a human activity—an activity of those who are both body and soul, who, though animals, are oriented toward the divine. (Indeed, one might take the discussion as an extended reflection on a sentence from the *Talmud* which Kass cites: "Whoever eats in the street or at any public place acts like a dog."[21]) To think about our need for food compels us to realize how we move from (animal) feeding, which would be sufficient if nourishment were the only desideratum, to truly human eating in friendly community, to ritual sanctification of the meal, pointing toward the transcendent. Thus, Kass's discussion of eating depicts it as activity of a creature who is animal, but whose animality is taken up and humanized in civilized dining and, in the end, perfected in religious ritual. To reflect upon eating is to think in this way about human dignity, about the complexity of a creature who is neither beast nor god. Pinker may be right, as I incline to think he is, on the limited question of public licking of ice-cream cones, but on the subject of human dignity he is tone deaf.

Finally, in his review Pinker offers three reasons why the notion of dignity is not well suited to play a foundational role in bioethics. Of these three, the second—that dignity is fungible—is germane here.[22] Pinker means that, while dignity may be of value to us, it is by no means the only thing we value. And we may sometimes—perhaps often—be willing to sacrifice the value of dignity in exchange for something else of value. "Doffing your belt and spread-eagling to allow a security guard to slide a wand up your crotch is undignified.... Most readers of this article have undergone a pelvic or rectal examination, and many have had the pleasure of a colonoscopy as well. We repeatedly vote with our feet (and other body parts) that dignity is a trivial value, well worth trading off for life, health, and safety."[23]

How little of what is most important in our concept of dignity such an example captures ought by now to be clear. To be sure, it captures something—roughly, what we mean by "dignified conduct." That notion is parasitic to some degree upon the more basic notion of *human* dignity, which suggests, inevitably, that only some of us and some of our behavior display humanity at its fullest and best. And though we might accept certain "indignities" (a pelvic or rectal examination for the sake of health), we understand that this involves a kind of temporary reductionism, abstracting the body from the person present in and through it. Indeed, we can find a scale on which to weigh these relative values only if we do for the moment treat the body more as a thing than as a place of personal presence. Moreover, there are limits to the "fungibility" we would accept in our concept of human dignity, though there is, as I have noted, no recipe book that can lay out these limits in formulaic manner. Discernment is needed. But, for example, the fact that I might value greatly another daughter to care for me in my old age would not lead me to clone a daughter I already have—as if I could trade off the disvalue that cloning involves (because it demeans human dignity) against the benefit of an additional daughter to care for me.

When we turn to what I have called personal dignity, the notion of comparing or weighing values has no place at all. Richard Stith has noted perceptively how inadequate is the language of value to capture our judgments about respect for personal dignity.[24] Thus, for example, if I have four children, I might place relatively little value on having another. Yet, if a fifth child is born to me, I would not consider killing him. His existence now requires of me not a comparative assessment of the value of his presence but, simply, respect. Similarly, to take an example from the end of life, I might be reluctant to spend thousands of dollars on a possibly life-prolonging operation but would not kill myself in order to save the same amount in estate taxes. To live is constantly to make choices among goods that we value, choices that will often involve judgments about the sorts of exchanges we are willing to make. But the dignity of the person, which gives each of us equal standing, is not something upon which we can place such a value. The language of personal dignity—because it deals in wrongs, not harms—is used to block just these sorts of exchanges. The dignity of each human person is to be respected, which is quite a different thing from being valued. A world of things we value, things that are fungible, is a world "subject to our evaluation and control." But respect, by contrast, "responds. It eschews control. . . . A limit is given to us and to our schemes of domination."[25]

Pinker's confusions are, therefore, instructive. To think them through is to see that there is good reason to examine aspects of human dignity, as I have in earlier chapters—and good reason, as well, to think yet about the dignity not of the species but of the person. To that last task we now turn.

CHAPTER EIGHT

Equal Persons

BETWEEN THE CONCEPTS of human dignity and personal dignity there is a dialectical relationship. Each needs the other to supplement its central concern. We need the language of human dignity to talk about matters that involve the integrity and flourishing of the human species, and we need the language of personal dignity to express respect for persons regarded as equal and non-interchangeable individuals.

The language of human dignity points to qualities that characterize and distinguish human life, and it will always be true that some individuals possess more of these qualities or possess them in more developed ways than do others. That is, some of us flourish more than others, showing in our lives what human beings at their best can be. Hence, the concept of human dignity invites comparative assessments, which suggest that some of us have greater dignity than others and that some may have lost dignity almost entirely.

Against the dangers of such comparison the notion of personal dignity provides protection. And, of course, it in turn, by affirming our equal dignity may seem to undermine all the distinctions of lesser or greater, better or worse, that we routinely make—and almost surely need to make—in many areas of life. Hence, the idea of human dignity is also needed to articulate what it means for human beings to flourish.

THE RISE TO PROMINENCE OF THE LANGUAGE OF DIGNITY

The idea of the equal dignity of every person—whatever the differences in our capacities, whatever our virtues or vices—is by no means a purely contemporary one. Although in the classical world of Greece and Rome dignity was generally an aristocratic concept—referring to those of high social status or those marked by excellence of one sort or another—we should not forget that Stoic thinkers did have an idea of a dignity that inhered in the individual abstracted from particular characteristics or social roles.[1]

The idea, though not necessarily the term, was certainly present in much Christian thought, which, whatever the failings in practice, taught an equal respect for each person as one made for covenant community with God and honored by the fact that in Jesus God had taken our humanity—even at its weakest and most vulnerable—into His own life. We can hardly overestimate the power of the Christmas story—of God as a baby in a manger—to shape our understanding of the dignity of every person. There is deep insight in the words of the French Christmas carol, "O Holy Night": "Long lay the world in sin and error pining, / Till He appeared, *and the soul felt its worth.*"[2] Thus, the story is told

> about the scholar Muretus, who in the year 1554 was ill, and the doctors proposed to try an operation on him. It was of the nature of an operation, but so slight were the chances of success, and so little their interest in healing him, compared with their desire to see what the symptoms would be before death came, that it would be fairer to call it an experiment in vivisection. Not knowing who the patient was, or that he spoke Latin, one doctor said to the other, '*Fiat experimentum in corpore vili*'

> ('Let the experiment be tried on this vile body'). '*Vilem animan appellas,*' came a voice from the bed, '*pro qua Christus non dedignatus est mori?*' ('Dost thou call that soul vile for which Christ was content to die?').[3]

The modern age has seen an increasing democratization of the idea of personal dignity even while, simultaneously, the ground of that equal dignity has become, in the minds of many, less obvious. When the view that grounds human dignity in our relation to God begins to recede in the minds of some thinkers, the theoretical case for equal respect must be made—if it can be made—on other grounds.

One way, roughly Kantian, is to focus on human freedom—and, in particular, on the freedom to choose for ourselves a way of life and prescribe for ourselves norms that will govern it. The great problem facing this approach though is that, in attempting to salvage personal dignity, it may lose the body and the human dignity of bodily life. Autonomous human beings characterized primarily and perhaps almost entirely by the freedom to shape and reshape themselves no longer must come to terms with purposes (or even a destiny) built into organic life. No longer simply made in the image of God, they are now free spirits, almost godlike themselves.

The other way, roughly Hobbesian, is to accent our shared vulnerability, which makes us approximately equal sharers in a life that threatens to be (as Hobbes famously described it) "solitary, poor, nasty, brutish, and short." But an emphasis upon that shared frailty may create as many problems as it solves. For, encouraging us to do whatever may be needed to relieve suffering, to sustain and extend life, or just to satisfy our desires, this ground for equal personal dignity is likely—as we have seen in earlier chapters—to undermine important aspects of human dignity.

As noted previously, the Universal Declaration of Human Rights, adopted by the United Nations in 1948, refers in its preamble to "the inherent dignity . . . of all members of the human family," and Article I states categorically that "all human beings are born free and equal in dignity and rights." The influence of the Declaration, though difficult to quantify, has surely been considerable, even though it offers no particular theoretical support for its assertion of human dignity. Hence, Mary Ann Glendon can suggest that the Declaration "succeeded well enough to give the lie to claims that peoples with drastically opposed worldviews cannot agree upon a few common standards of decency."[4]

I have some doubts, to which I will come shortly, about whether such an approach can really succeed, but, if it fails, the failure is a noble one. In the process of preparing the Declaration, UNESCO convened a committee of philosophers to examine the theoretical bases for universal claims about human dignity and rights. While the philosophers were able to agree on many particular claims, they were, perhaps unsurprisingly, unable to agree on "why" these claims are true—unable, that is, to develop any shared vision of human nature or the human person on which such claims could be based. Jacques Maritain, one of the participating philosophers, later recounted how

> at one of the meetings of a UNESCO National Commission where human rights were being discussed, someone expressed astonishment that certain champions of violently opposed ideologies had agreed on a list of those rights. "Yes," they said, "we agree about the rights *but on condition that no one asks us why*."[5]

Surely, there is something to this. Our most fundamental moral convictions, precisely because they are fundamental, cannot be deduced from or proven by any more basic moral truths. They must shine by their own light. We can argue *from* them—indeed, without them we would hardly know how to carry out moral argument and discussion—but not *to* them. "It is," as C. S. Lewis wrote, "no use trying to 'see through' first principles. . . . To 'see through' all things is the same as not to see."[6]

Although first principles must shine by their own light, that light sometimes seems dim or clouded, and agreement is not readily achievable. And, in fact, appeals to human dignity, such as that in the

Universal Declaration, have often been criticized as a kind of moral colonialism, an exporting of fundamentally Western beliefs into other cultures where they are alien. While there is something to this, we could easily overstate it. Our very ability to engage in discussion about important moral questions with those whose starting points are quite different from our own—the analogy in moral argument to our ability to translate from one language into another—suggests that some common ground is shared. Thus, Mary Ann Glendon argues that "what was crucial" for the framers of the Universal Declaration, who themselves faced deep cultural and ideological divides, was "the *similarity* among all human beings. Their starting point was the simple fact of the common humanity shared by every man, woman, and child on earth, a fact that, for them, put linguistic, racial, religious, and other differences into their proper perspective."[7]

Of course, that very conviction depends on a belief that "being human" is more fundamental than the differing capacities and perspectives that distinguish human beings from each other, and the nominalism so commonly assumed in our day is often more impressed by individual differences than by a shared humanity. This becomes especially evident when we ask who belongs, who gets counted morally, how inclusive—or how restricted—is the meaning given to "all human beings." We have to wonder, therefore, what entitles us to be confident that, even apart from any shared metaphysic, we will in practice agree that some deeds, at least, are so evil that they cannot be defended.

I suspect that what gives rise to such confidence is a belief so basic that we seldom articulate it—that, as Augustine says, "God created man as one individual . . . that in this way the unity of human society and the bonds of human sympathy be more emphatically brought home to man."[8] That is, even Glendon's invocation of "the simple fact of common humanity" as morally significant may be nourished by background beliefs whose roots are religious. Indeed, it may be that such an *epistemologically particular* starting point gives the surest ground for a confidence in *ontologically universal* claims about the dignity of every human being. Thus, for example, David Novak suggests that one can see in Abraham Joshua Heschel's participation in the civil rights movement how the particularity of Jewish revelation undergirded a "universality inherent in Jewish ethics." For Heschel, public humiliation was a form of oppression that violated the rabbinic principle of human dignity (*kevod ha-beriyot*), which permitted the overriding even of certain religious restrictions when they "would lead to the public humiliation of another person."[9]

Because every person is the bearer of an equal dignity, and because there are such universal moral truths, we can and should have a kind of confidence that—however these truths may for a time be ignored or violated—they will sooner or later reassert their claim upon us. But, of course, it may be later rather than sooner, and—as was the case with the appeal to the inherent dignity of every person in the Universal Declaration—this reassertion may come only after we have witnessed and sometimes perpetrated terrible evils. Moreover, as I noted earlier, even when we affirm the equal dignity of every person, we must somehow make that affirmation of equality cohere with the other truth of human dignity: that even as equal persons we do, in fact, differ considerably in the degree to which we exemplify the dignity of being human.

THE DIGNITY OF THE PERSON

It is obvious that, at least in certain contexts and for certain purposes, we make distinctions of merit among human beings. Academic institutions, for example, are meritocratic, and a class in which every student gets an A—even if welcomed for certain reasons by some students and some faculty—is understood to subvert the very nature of the undertaking. Likewise, the worlds of sport and of musical performance—to take two quite different aspects of life—are arenas in which we still strive for excellence and watch with an eye to discerning those whose performance is especially accomplished. We generally think that an eye for these distinctions and differences need not undercut our commitment to the equal dignity of human beings. Perhaps it need not, but we must keep in mind the difference between thinking about the dignity characteristic of the human species and about the dignity inhering in each person.

There are circumstances in which we tend quite naturally to think in terms of such a distinction. Suppose, for example, a cloned human being were born, a possibility we have had to contemplate at least since the cloning of the sheep Dolly, announced in 1997. Many people, whether they can articulate their reasons or not, think such an event would demean or degrade human dignity. (If developed, their reasons would probably be similar to my earlier discussion in the chapter on "birth and breeding.") Yet, many of these same people, if asked whether the newly cloned person deserves respect as our equal, or is really one of us, or (as the point is sometimes put) has a soul, would not hesitate to answer yes. Thus, to take an example that was not at all unusual at the time, in an article discussing these issues shortly after Dolly's birth, a theologian was quoted as follows: "I would say that a cloned human would be as much a child of God as a non-cloned human."[10] One can quite coherently believe both that cloning would undermine the dignity that characterizes our humanity and that a cloned human being would be our equal in the dignity that inheres in every person.

A very different sort of example will make the same point. Discussing the morality of capital punishment, constitutional scholar Walter Berns quotes Supreme Court Justice William Brennan's statement that "even the vilest criminal remains a human being possessed of human dignity"—and then disagrees emphatically.

> What sort of humanism is it that respects equally the life of Thomas Jefferson and Charles Manson, Abraham Lincoln and Adolf Eichmann, Martin Luther King and James Earl Ray? To say that these men, some great and some unspeakably vile, equally possess human dignity is to demonstrate an inability to make a moral judgment derived from or based on the idea of human dignity.[11]

We understand what Berns means, and in certain moods we are probably inclined to agree; yet, in my view, the more striking inability displayed in this passage is Berns's own inability to find a standpoint from which to see the whole truth about any and every human life. Especially when life and death are at stake, when we are forced to think about a person's life as a whole, the distinctions that we make and need to make in other contexts may lose their force. To continue to respect the dignity of a person who has done great evil, who may have diminished or demeaned the human dignity he shares with us, is no easy task. Yet, our very willingness to attempt to give that person a fair trial, and—puzzling as it may seem sometimes—our willingness to hold him responsible and punish him depend on continued affirmation of the dignity of his person.

Likewise, all of us have probably known people who hold—and perhaps, even, vigorously assert and defend—views that we find morally abhorrent, views that may seem to us to be assaults on human dignity. Sometimes, in fact, such people may come to occupy places of prominence within society, particularly in the academy. And we may engage with them, as best we can, in civil argument and discussion. Although we might think our society would be healthier if their views were simply eliminated from public discussion, we would not, even if we could, eliminate *them*. For, however repugnant their views may be, the one who articulates those views is also a bearer of personal dignity, who can never be thought of apart from the relation to God, which we all share equally.

Were human beings simply members of their species, were that human dignity the only dignity they bear, it might sometimes make sense to sacrifice one or another of us for the sake of the species as a whole. If, however much we value the distinctive features of our humanity, we think that would be wrong, it is because a person is a "someone," not a "something." A person not only shares in the value of the species but also "occupies a unique and distinctive position entirely his or her own" that transcends species membership.[12] Thus, though all human beings share in human dignity, they are not interchangeable. "The *value* of ten people may be more than that of one, but ten are no more than one in point of [personal] dignity."[13]

Suppose, Kierkegaard writes,

> there are two artists and one of them says, "I have traveled much and seen much in the world, but I have sought in vain to find a person worth painting. I have found no face that was the perfect image of beauty to such a degree that I could decide to sketch it; in every face I have seen one or another little defect, and therefore I seek in vain." Would this be a sign that this artist is a great artist? The other artist, however, says, "Well, I do not actually profess to be an artist; I have not traveled abroad either but stay at home with the little circle of people who are closest to me, since I have not found one single face to be so insignificant or so faulted that I still could not discern a more beautiful side and discover something transfigured in it. That is why, without claiming to be an artist, I am happy in the art I practice and find it satisfying." Would this not be a sign that he is indeed the artist, he who by bringing a certain something with him found right on the spot what the well-traveled artist did not find anywhere in the world—perhaps because he did not bring a certain something with him! Therefore the second of the two would be the artist.[14]

The equal dignity of persons may not always seem obvious, of course. Indeed, perhaps we will see it only insofar as we "bring a certain something" with us when we look. And, for Kierkegaard, that "certain something" is very specifically the neighbor-love that Christians are enjoined to show to every human being made for covenant community with God. I doubt, in fact, that there is any way to derive a commitment to equal respect for every human being from the ordinary distinctions in merit and excellence that we all use in some spheres of life; it is grounded, rather, not in our relation to each other but in our relation to God, from Whom—to return to Kierkegaard's mathematical metaphor—we are equidistant. "The thought of God's presence makes a person modest in relation to another person, because the presence of God makes the two essentially equal."[15]

Here, then, is our problem, from which we cannot for long continue to avert our gaze: Our society is committed to equal dignity, and our history is in large part a long attempt to work out the meaning of that commitment. Christians and Jews have an account of persons—as equidistant from God and of equal worth before God—that grounds and makes sense of this commitment we all share. A society that rejects their account but wishes to retain the commitment faces, then, a serious crisis in the structure of its beliefs. And often, in fact, we do little more than posit an equal dignity about which we are, otherwise, largely mute; for the truth is, as Oliver O'Donovan has assertively put it, that this belief "is, and can only be, a theological assertion."[16] We are equal to each other, whatever our differences in human flourishing, precisely because none of us is the "maker" of another one of us. We have all received our life—equally—as a gift from the Creator.

This does not mean that equal personal dignity can or will be affirmed only by religious believers. In developing a roughly Kantian account of what it means to respect the life and dignity of each person, the philosopher J. David Velleman grants that "talk of someone's value as a person sounds like religion rather than philosophy."[17] He suggests that "our values will be incoherent so long as they lack a counterpart to the sanctity of human life."[18] Still, without fully discerning the ontological ground of dignity one may have what Gabriel Marcel terms "an active and even poignant experience of the mystery inherent in the human condition."[19] We will gain insight into this mystery chiefly, Marcel thinks, when we are moved by a spirit of compassion that recognizes our shared vulnerability; hence, "dignity must be sought at the antipodes of pretension and . . . on the side of weakness."[20] That is to say, in our common subjection to mortality—to death, in which we must discern the meaning of a life taken whole—we may come to perceive dimly our equal dignity.

At any rate, it is not religious believers who should be ill at ease in a public square committed to equal respect for every human being; it is those who lack the faith that animated and animates such commitment. It is not religious believers who should be mute in a public square committed to equal dignity; it is others who find themselves mute when asked to give an account of our shared public commitment. In fact, an appreciation of the many and various distinctions in human flourishing—of the sort embedded in the concept of human dignity—is safe only in a public square that can affirm the relation to the Creator which grounds our equality as persons.

Thus, we can grant and make use of comparative assessments of value (which arise inevitably from the idea of human dignity) as long as that use is shaped and transformed by commitment to a non-comparative and equal dignity of persons. This shaping will show itself and be important in at least two ways.[21] First, it may enable us to see what we otherwise might not were we to look only at surface differences—even *important* surface differences. It will form us as people rather like Kierkegaard's second artist, whose eye is attuned to the deeper truth that lies behind, beneath, and within the differences that distinguish us from each other.

In addition, this non-comparative concept of dignity will become relevant whenever we make what we might call "on the whole" judgments about the worth of a human life. Unable to transcend entirely our location in time and space, we never see any life, including our own, in such a transcendent way. It presupposes, really, God's own perspective; hence, in making such judgments we think of ourselves and others in terms of the relation to God. This need not blind us to the many distinctions within everyday social life, for dissimilarity is, as Kierkegaard notes, the mark (though a confusing mark) of temporal life. "But the neighbor is eternity's mark—on every human being."[22] Since we stand equally distant from (or near to) the Eternal One, we are radically equal in those moments when our life is judged "on the whole," as only God can see it.

One place, therefore, where differences in excellence or dignity can have no place, will be at "the threshold of death, when the continuance of life itself is at stake."[23] Eugen Rosenstock-Huessy recounts an old ritual in Austria in which

> the corpse of the emperor was ordered to be carried to the door of an abbey. The chamberlain who leads the cortège knocks at the door. A friar opens the window and asks: "Who knocks?"—"The Emperor."—"I know no man of that name." The chamberlain knocks again. "Who is there?"—"The Emperor Francis Joseph."— "We do not know him." Third knock, and the same question. After reflection, the chamberlain now answers: "Brother Francis." Then the door opens to receive a comrade in the army of death, on equal terms with all souls."[24]

Once again, Kierkegaard sees the point: "There is not a single person in the whole world who is as surely and as easily recognized as the neighbor. You can never confuse him with anyone else, since the neighbor, to be sure, is all people. . . . If you save a person's life in the dark, thinking that it is your friend—but it was the neighbor—this is no mistake."[25]

We also encounter others "on the whole," (and differences in excellence become unimportant) when "they lack essential resources to participate in social communications as such."[26] Every human being—created by God for covenant with each other and with himself, even in the midst of the many distinctions that mark us—must have the opportunity to live within human society and participate in its common life. Thus, "the opportunity to live, and the opportunity to participate in a society, are metaphysically foundational; they correspond to our universal created nature as human beings."[27] Recognizing these two forms of "on the whole" equality need not efface our appreciation for the significance of differences among us in excellence and achievement, but it will inevitably, I suspect, democratize somewhat the judgments we make about the worth of human lives. Even within our noblest qualities and our most striking excellences, we will learn to discern "the poverty of our perfections."[28]

PERSONAL DIGNITY AND DISTINCTIONS IN HUMAN EXCELLENCE

The difficulty of holding together our belief in the equal dignity of persons and our sense that the dignity of human life is displayed more in some circumstances than in others has become especially apparent in our time in problems of bioethics. We can observe the President's Council on Bioethics wrestling with one instance of such difficulties in its 2005 report *Taking Care: Ethical Caregiving in Our Aging Society.*

A distinction between two different senses in which one might speak of human dignity is specifically noted in the report. For example, the language of dignity may be used to mark either a "floor," a kind of respect and care beneath which our treatment of any human being should never fall—or it might be used to mark a "height" of human excellence, those qualities that distinguish some of us from others.[29] Similarly, the report notes a difference between an "'ethic of equality' (valuing all human beings in light of their common humanity)" and an "'ethic of quality' (valuing life when it embodies certain humanly fitting characteristics or enables certain humanly satisfying experiences)."[30]

The general point is, I think, clear, and it seems right to say that, at different times and for different purposes, we are likely to speak in either of these ways. Nonetheless, trying to find a way to do justice to each of them simultaneously is no easy task. Nor am I persuaded that the Council's discussion is entirely successful, for it seldom does more than set the two concepts of dignity side by side. They do not interact in such a way that the meaning of one can be to some degree reshaped or transformed by the other; instead, they remain firmly fixed in separate linguistic compartments. For example, having discussed a (comparative) sense in which we might think of some human beings as manifesting greater dignity than others, the Council then turns to affirm a "*non*-comparative . . . way of speaking about the worth of human lives."[31] Yet, attempting to affirm this non-comparative worth, it says merely: "If we value *only* the great ones, we do an injustice to the dignity of ordinary human beings."[32]

Suppose, however, that our understanding of comparative excellence, of differences in the degree to which different people's lives display various aspects of human dignity, were reshaped somewhat by a sense of respect for the equal dignity of persons. We might then incline to draw back a bit from some elements in the Council's discussion. For example, imagining a woman who was once a "virtuoso violinist" and is now suffering from dementia, her "treasured capacities" largely gone, the Council first affirms that she "remains a full member of the human community, equally worthy of human care." But it then expresses puzzlement about what her dignity might mean when those capacities are "fading or gone." In the case of such a virtuoso, the suggestion seems to be, dementia is especially degrading. "For all people—and perhaps most vividly for those who once stood high above the ordinary—the regression to dementia and incompetence, with all its accompanying indignities and loss of self-command, may seem dehumanizing and humiliating."[33]

We should, I think, be hesitant to speak this way. I cannot see why dementia afflicting this "virtuoso violinist" should be any more vividly dehumanizing than it would be were it to afflict, say, the woman who regularly empties the trash can in my office. Still more, I would be reluctant, without considerable qualification and explanation, to call dementia in either case dehumanizing. It does not subvert the dignity of birth, breeding, or death. It does, of course, diminish and attack certain characteristic—and valuable—human capacities, making it impossible for us to continue to flourish. But frailty and decline are part of being human; in all of us the fires of metabolism eventually die down.

I am reluctant to say that any living human being, even one

severely disabled by dementia, has lost the dignity that belongs to all human persons. Why? I am reluctant to say that some persons—those with certain highly developed capacities—have greater dignity than others. Why?

These two puzzles are interrelated, and I suspect that my reluctances are grounded in a view that permits our understanding of human dignity to be subordinated to and—to some extent—transformed by our understanding of the equal dignity of persons. If we assert that every human being has dignity, someone is certain to ask from us an account of what it is about human beings that gives them this equal dignity. And of almost every characteristic or property to which we might point it is likely that some human beings may lack it or lose it, or that some human beings may have it in more developed or more excellent ways (and, hence, may seem more worthy or of greater value). If dementia is inherently dehumanizing because it deprives human beings of the rational powers that make them special, then some living human beings may come to lack dignity entirely. If dementia is worse when it attacks the "virtuoso," diminishing qualities that were once especially highly developed, it would seem that the virtuoso and the janitor were never of equal dignity.

We can deal safely with these puzzles and difficulties only when we put personal dignity first. A society can acknowledge and reward differences in accomplishment and achievement, it can recognize the sadness and tragedy of disability and fading capacities, and it can appreciate the worth of particular loves and special bonds of association—it can, that is, honor and affirm the dignity of the human condition, of this creature who is neither beast nor god. But it can safely do this only when its first and last commitment is to respect the equal dignity of persons, each of whom is made for community with God. It will sometimes be difficult to sort out the relation between these two concepts of dignity, and it is no shame to find ourselves sometimes puzzled and uncertain, but it is an effort that honors the human dignity we share.

THE COINHERENCE OF HUMAN AND PERSONAL DIGNITY

To be a person is not to *have something* but to *be someone.* It is persons who are equidistant from Eternity and equal in dignity. What is this but to say that we today sometimes get these matters exactly turned around? For we now often speak of "personhood" as something a human being may—or may not—have. The category of personhood is used to distinguish some human beings from others, to deprive some of the dignity of persons. It is used to deny rather than to affirm our fundamental equality.

It should be obvious, I hope, that I have turned in quite a different direction. We all share in the dignity of being human and should do our best to honor and uphold the characteristic human form of coming into being, living in community with others, and dying. But we are not just members of a species or instances of a universal type, and the nature that we have must be distinguished from the person who we are. "The point of a proper name," Ralph McInerny notes,

> is that it [is] not common to many, and yet many people do bear identical names. . . . But even when two persons have the same proper name it does not become a common noun, like 'man.' All the John Smiths that have been, are, and will be have nothing in common but the name; it does not name something common to them all. There is an inescapable nominalism here. God calls us all by our proper name, and He is unlikely to confuse one John Smith with another.[34]

As human beings we share the characteristic human form and participate in its dignity, whatever our individual traits or capacities; as persons who always exist in relation to God, we are radically individual and equal. Still, what we can and must separate in thought is not separate in our lived experience.

To be a person is to be characterized by a mysterious inwardness, an inner distance from oneself, which we experience but find hard to articulate. A stone falls from a building and is simply an object constrained by the laws of nature. I fall from that same building and know myself as a falling object—which I both am and am distanced from. This is, as we have seen, inherent in the human being's mediated relation to the surrounding world. Each of us is both human being and person, and we experience this inner distance, this embodiment and transcendence of the body, in everything we do.

But this connection of body and person, so taken for granted in our own experience, is crucial for our coming to know other persons and respecting their dignity. The body is the place of their personal presence. We know them only there, as they likewise know us only as embodied persons. In knowing them this way, however, we do not assess their personal dignity—or their equality with us—on the basis of the presence or absence of various characteristically human capacities. Rather, we "come to know" them, we recognize them as persons—which is quite a different thing from responding to or assessing certain personal properties. Hence, we need not ignore their differences or fear that noting these differences will undermine our equal dignity, for that dignity is known not by observation or assessment but, rather, by recognition. "The mother, or her substitute, treats the child from the start as a subject of personal encounter rather than an object to manipulate or a living organism to condition. She teaches the child to speak, not by speaking in its presence but by speaking *to* it."[35] She does not first cultivate certain qualities in a little something until, at some point, that something becomes someone.

The dignity of our humanity and the dignity of our person thus coinhere. We know persons only as bodies, and when we encounter a living human body our moral task is to seek to recognize the person who is there. I doubt that anyone can simply be compelled into such recognition by rational argument alone. The heart must be open to recognize personal dignity in every living human being (even those who are far from flourishing in the ways that characterize the dignity of the human species). We must be ready to set aside the notion that we should evaluate their claim to personal dignity and accept the truth that, in our willingness or unwillingness to acknowledge it, we judge ourselves.

NOTES

CHAPTER ONE · SPEAKING OF DIGNITY

1 Søren Kierkegaard, *The Point of View for My Work as an Author*, trans. Walter Lowrie (New York: Harper & Brothers, 1962), 124.
2 St. Augustine, *City of God*, 12.22.
3 C. S. Lewis, *That Hideous Strength* (New York: Macmillan, 1965), 174.
4 *Being Human. Readings from the President's Council on Bioethics* (New York and London: W.W. Norton, 2004), 567.
5 Hans S. Reinders, "Human Dignity in the Absence of Agency," in R. Kendall Soulen and Linda Woodhead (ed.), *God and Human Dignity* (Grand Rapids, MI: Eerdmans, 2006), 122–23.
6 Peter Singer, "A Convenient Truth," *The New York Times* (January 26, 2007).
7 Søren Kierkegaard, *Purity of Heart Is To Will One Thing*, trans. Douglas V. Steere (New York: Harper & Row, 1948), 191.
8 Kierkegaard, *Purity of Heart*, 208. See G. K. Chesterton, *Autobiography* (London: Hutchinson & Co., 1937), 332: "It is not familiarity but comparison that breeds contempt."
9 Thomas Aquinas, *Summa Theologiæ*, IIaIIae, q. 64, a. 2, ad. 3. We should not imagine that this idea is a peculiarity of Aquinas. Thus, for example, in his Second Treatise, John Locke writes (par. 11) that a murderer may be killed in order "*to secure* Men from the attempts of a Criminal, who having renounced Reason, the common Rule and Measure, God hath given to Mankind, hath by the unjust Violence and Slaughter he hath committed upon one, declared War against all Mankind, and therefore may be destroyed as a *Lyon* or a *Tyger*, one of those wild Savage Beasts, with whom Men can have no Society nor Security."
10 John Paul II, *The Gospel of Life* (Boston, Pauline Books & Media), par. 9. The words quoted here are actually italicized in the English translation, which inserts "personal" in a rather free translation of the Latin.

CHAPTER TWO · BEING HUMAN

1 Aristotle, *Nicomachean Ethics*, 1097b.
2 *Ibid.*, 1098a.
3 *Ibid.*
4 Hans Jonas, *The Phenomenon of Life: Toward a Philosophical Biology* (New York:

Harper & Row, 1966); *Philosophical Essays: From Ancient Creed to Technological Man* (Englewood Cliffs, NJ: Prentice-Hall, 1974).

5 Jonas, *The Phenomenon of Life*, 75–76.

6 Hans Jonas, "The Burden and Blessing of Mortality," *Hastings Center Report*, vol. 22, no. 1 (January–February, 1992), 35. Reprinted in *Mortality and Morality: A Search for Good After Auschwitz* (Chicago: Northwestern University Press, 1996), 89.

7 We can see here a connection with Jonas's earlier important work on Gnosticism. The duality of inner self and external world that is inherent in organic life can easily be transformed into a dualism that separates, rather than just distinguishes, spirit from matter. Jonas himself noted these connections in "Wissenschaft as Personal Experience," *Hastings Center Report*, vol. 32, no. 4 (July–August, 2002): "The very possibility of such dichotomies says something about the human being, about us. . . . It was my long involvement with dualism that was of particular benefit to me in re-examining the German field of the philosophy of consciousness in which I had been trained" (31). And again: "My ontological interpretation of the organism was intended to correct this error [dualism]. . . . In organic being's essential unity of 'inner' and 'outer,' subjectivity and objectivity, free self and causally determined thing, the gulf between matter and mind closed for me" (33).

8 Jonas, *The Phenomenon of Life*, 76.

9 Jonas, *Philosophical Essays*, 193.

10 Aquinas, IaIIae, q. 94, a. 2.

11 Jonas, *The Phenomenon of Life*, 80.

12 Jonas, *Philosophical Essays*, 204.

13 Jonas, *The Phenomenon of Life*, 112.

14 *Ibid.*, 122.

15 Leon R. Kass, "Appreciating The Phenomenon of Life," *Hastings Center Report*, vol. 25 (Special Issue, 1995), 7.

16 Jonas, *Philosophical Essays*, 205.

17 Kass, 11.

18 Aquinas, IaIIae, q. 94, a. 2.

19 Karl Barth, *Church Dogmatics*, III/2 (Edinburgh: T. & T. Clark, 1960), 294.

20 Kass, 11.

21 C. S. Lewis, *An Experiment in Criticism* (Cambridge: the University Press, 1969), 138.

22 See Søren Kierkegaard, *The Point of View for My Work as an Author*, trans. Walter Lowrie. (New York: Harper & Brothers, 1962), note, 88–89: "I have endeavoured to express the thought that to employ the category 'race' to indicate what it is to be a man, and especially as an indication of the highest attainment, is a misunderstanding and mere paganism, because the race, mankind, differs from an animal race not merely by its general superiority as a race, but by the human characteris-

tic that every single individual within the race (not merely distinguished individuals but every individual) is more than the race. This follows from the relation of the individual to God."

23 Aquinas, IaIIae, q. 94, a. 2. Perhaps this does not really press so far beyond Jonas's philosophical biology, for his analysis of the "nobility of sight" discovered in the possibility of a visual presentation that "has no intrinsic reference to time" the germ of the idea of eternity and the distinction between being and becoming (*The Phenomenon of Life*, 142). It is sight that gives us "the present as something more than the point—experience of the passing now" (144).

CHAPTER THREE · BIRTH AND BREEDING

1 Leon Kass, Francis Fukuyama, Bill McKibben, and Jeremy Rifkin.

2 Nicholas Agar, "Whereto Transhumanism?," *Hastings Center Report*, vol. 37, no. 3 (May–June, 2007), 12.

3 Though we should be careful with this formulation, since there is a god—the One Whom Christians believe to have been revealed in Jesus—Who knows through experience both suffering and death.

4 William Godwin, *An Enquiry Concerning Political Justice and Its Influence on General Virtue and Happiness*, vol. 1. Abridged and edited from the 1793 edition by Raymond A. Preston (New York: Alfred A. Knopf, 1926), 42.

5 C. S. Lewis, *Miracles* (New York: Macmillan, 1947), 166.

6 Thomas Hobbes, *The Citizen: Philosophical Rudiments Concerning Government and Society, in Man and the Citizen*, ed. Bernard Gert (Garden City, NY: Doubleday Anchor, 1972), IX.1.

7 Thomas Hobbes, *Leviathan*, ed. Michael Oakeshott (New York: Collier Books, 1962), 155.

8 A. I. Melden, *Rights and Right Conduct* (Oxford: Basil Blackwell, 1959), 26.

9 Ronald Bailey, *Liberation Biology: The Scientific and Moral Case for the Biotech Revolution* (Amherst, NY: Prometheus Books, 2005), chapter 5.

10 Paul Ramsey, "Shall We 'Reproduce'?: I. The Medical Ethics of In Vitro Fertilization," *The Journal of the American Medical Association*, 220 (June 5, 1972), 1482.

11 Rebecca Bennett and John Harris, "Reproductive Choice," in *The Blackwell Guide to Medical Ethics*, ed. Rosamond Rhodes, Leslie P. Francis, and Anita Silvers (Oxford: Blackwell Publishing, 2007), 201.

12 Bennett and Harris, 201.

13 Psalm 8:5.

14 Kurt Bayerz, "Human Dignity: Philosophical Origin and Scientific Erosion of an Idea," in *Sanctity of Life and Human Dignity*, ed. Kurt Bayertz (Dordrecht, Boston, London: Kluwer, 1996), 76.

15 *Ibid.*, 77.

16 Bennett and Harris, 202.

17 John A. Robertson, "Procreative Liberty in the Era of Genomics," *American Journal of Law & Medicine*, vol. 29 (2003), 439–487. In explicating and examining Robertson's position, I will regularly write of "reproductive" rather than "procreative" liberty. That is because I do not understand him to be advocating anything that should be called procreation. What he has in mind is fundamentally an act of manufacture, of making. But, to put to use a very ancient distinction, procreation is an act of doing rather than making.

18 *Ibid.*, 445.

19 *Ibid.*, 444.

20 Paul Ramsey, *Fabricated Man: The Ethics of Genetic Control* (New Haven and London: Yale University Press, 1970), 36.

21 See Brent Waters, *Reproductive Technology* (Cleveland: Pilgrim Press, 2001), 18: For Robertson a "parent" is "one planning and managing a reproductive project from inception to completion."

22 Robertson, 450.

23 *Ibid.* A haploid is a set of chromosomes containing only one member of each pair, the genetic material in a sperm cell or an egg.

24 *Ibid.*, 452.

25 *Ibid.*, 451.

26 *Ibid.*, 450.

27 Kay S. Hymowitz, "The Incredible Shrinking Father," *City Journal*, vol. 17, no. 2 (Spring 2007).

28 Ryan T. Anderson and Christopher Tollefsen, "Biotech Enhancement and Natural Law," *The New Atlantis*, no. 20 (Spring 2008), 90.

29 Like most others, Robertson actually calls this selection for "gender." It is, however, impossible to select for gender, which is a social construct. One can only select for sex, which is a biological category.

30 Robertson, 471.

31 *Ibid.*, 480.

32 *Ibid.*

33 Bennett and Harris, 209. On ectogenesis and male pregnancy, see 217.

34 Ramsey, "Shall We 'Reproduce'?," 1481.

CHAPTER FOUR · CHILDHOOD

1 St. Augustine, *Confessions*, 1.1. Although I have modified it in this passage, I will generally quote from Rex Warner's translation (New American Library, 1963).

2 *Ibid.*, 1.13.

3 *Ibid.*, 1.19.

4 *Ibid.*, 2.1.

5 *Ibid.*, 2.4.

6 *Ibid.*, 2.10.

7 *Ibid.*, 3.1.
8 *Ibid.*
9 *Ibid.*, 6.6.
10 *Ibid.*, 9.2.
11 Peter Brown, *Augustine of Hippo: A Biography.* A New Edition with an Epilogue (Berkeley and Los Angeles: University of California Press, 2000), 102.
12 Augustine, 8.5.
13 *Ibid.*
14 *Ibid.*, 2.2.
15 *Ibid.*, 3.1.
16 *Ibid.*, 6.6.
17 *Ibid.*, 10.40.
18 The transcript is available at: www.bioethics.gov/transcripts/feb06/session5.html.
19 Augustine, 4.7.
20 *Beyond Therapy: Biotechnology and the Pursuit of Happiness.* A Report of the President's Council on Bioethics (New York: Regan Books, 2003), 76. Also available online at http://www.bioethics.gov/reports/beyondtherapy/index.html.
21 *Ibid.*, 75.
22 *Ibid.*, 94. A few dissenting opinions: Wilson Carey McWilliams, *The Idea of Fraternity in America* (University of California Press, 1973), 36: "The impression that childhood is a dulcet time suggests how kind are the failures of memory." George Orwell, "Such, Such Were the Joys. . . ," *A Collection of Essays* (Harcourt Brace Jovanovich, 1953), 4–5: He writes of "a deeper grief which is peculiar to childhood and not easy to convey: a sense of desolate loneliness and helplessness, of being locked up not only in a hostile world but in a world of good and evil where the rules were such that it was actually not possible for me to keep them." To which I might add the wisdom of my mother who not only raised her own five children but cared at various times for a number of other children as well, and who often said: "Children have problems of their own."
23 Gareth B. Matthews, *The Philosophy of Childhood* (Harvard University Press, 1994).
24 Gareth B. Matthews, "A Philosophy of Childhood." A monograph published by the Poynter Center for the Study of Ethics and American Institutions, Indiana University (January, 2006), 7.
25 Matthews, "A Philosophy of Childhood," 7.
26 *Beyond Therapy*, 73.
27 *Ibid.*
28 It was Eric Cohen who reminded me that these contrasting but also complementary images had first played a role in the Council's deliberations about the end of life, deliberations that reached their fruition in the 2005 report, *Taking Care: Ethical Caregiving in Our Aging Society.*
29 William H. Willimon, "Hard Truths: William H. Willimon on Paul Holmer," *The*

Christian Century, no. 122 (Feb. 22, 2005), 27–28.

30 Marilynne Robinson, *Gilead* (New York: Farrar, Straus, and Giroux, 2004), 129.

31 The transcripts are at: www.bioethics.gov/transcripts/dec02/session3.html (for Diller) and www.bioethics.gov/transcripts/march03/session1.html (for Hyman).

32 For Biederman see note 18 above. For the Eides: www.bioethics.gov/transcripts/feb06/session4.html.

33 See Peter Conrad, *The Medicalization of Society* (Baltimore: The Johns Hopkins University Press, 2007), 64: "What is interesting about adult ADHD is that many of those who are given the diagnosis are by some measures successful individuals."

34 *Beyond Therapy*, 91.

35 John Wisdom, *Paradox and Discovery* (Oxford: Basil Blackwell, 1965), 138.

CHAPTER FIVE · LOYALTIES

1 Acts 17:26.

2 Hosea 11:1.

3 Deuteronomy 29:14–15.

4 For an old but still fascinating discussion of the convoluted history of interpretation of this passage by Christians sensing the tension between it and what they believed to be the universal mission of the church, see Benjamin Nelson, *The Idea of Usury: From Tribal Brotherhood to Universal Otherhood.* Second Edition Enlarged (Chicago and London: The University of Chicago Press, 1969). This is clearly a book in which the action takes place in the subtitle.

5 Deuteronomy 10:14–19.

6 Amos 9:7.

7 Deuteronomy 2:5.

8 Isaiah 49:6.

9 Malachi 1:11.

10 Romans 1:16.

11 Luke 24:47; Acts 1:8.

12 Luke 10:25–37.

13 John 4:23.

14 John 1:14–17.

15 Revelation 21:1–4.

16 St. Augustine, *City of God*, 15.1.

17 Martha C. Nussbaum, "Patriotism and Cosmopolitanism," in *For Love of Country: Debating the Limits of Patriotism*, ed. Joshua Cohen (Boston: Beacon Press, 1996), 7. In context, Nussbaum is here discussing Stoic ideas, but they seem to be ideas she wishes to affirm in her essay.

18 *Ibid.*, 11.

19 *Ibid.*, 14.
20 *Ibid.*, 9.
21 *Ibid.*, 13.
22 I am focusing here on only one kind of problem raised by special moral relationships—namely, that they may seem to narrow the scope of our moral responsibility in ways that are unfair to those who are not our fellow "insiders." These particular bonds also raise another kind of problem, which, though important, is beyond the scope of my discussion here. In at least many cases—family being the most obvious, but nation perhaps the most problematic to many—they are bonds which we do not seem to have chosen for ourselves. Although I do not take up this concern here, it should be obvious that from my perspective any angle of vision that makes autonomy so central as to call into question all duties arising from unchosen human relationships is another attempt to escape the limits of our finitude and to be pure spirit. Samuel Scheffler is right to say that the idea that "our own wills are the source of all our special responsibilities" is a "fantasy" that we need to surrender. Samuel Scheffler, *Boundaries and Allegiances: Problems of Justice and Responsibility in Liberal Thought* (Oxford University Press, 2001), 107.
23 Augustine, 12.22.
24 Scheffler, 109.
25 Alasdair MacIntyre, "Is Patriotism A Virtue?" in *Theorizing Citizenship*, ed. Ronald Beiner (SUNY Press, 1995), 218.
26 Mark 3:35.
27 Karl Barth, *Church Dogmatics*, III/4 (Edinburgh: T. & T. Clark, 1961), 261.
28 Robert Spaemann, *Persons: The Difference between 'Someone' and 'Something'* (Oxford: Oxford University Press, 2006), 99. I have made this argument in more detail in chapter 5 of *Faith and Faithfulness: Basic Themes in Christian Ethics* (Notre Dame, IN: University of Notre Dame Press, 1991).
29 MacIntyre, 226.
30 C. S. Lewis, *The Four Loves* (New York: Harcourt Brace Jovanovich, 1960), 48.
31 Scheffler, 64.
32 *Ibid.*, 108.
33 Lewis, 41.
34 Augustine, 1.29.

CHAPTER SIX · DEATH

1 Homer, *The Odyssey*, trans. Robert Fagles (New York: Penguin Books, 1996), 1.12–18.
2 *Ibid.*, 5.233–234.
3 *Ibid.*, 5.236–243.

4 Sullivan Ballou, "Letter to Sarah," in *Wing to Wing, Oar to Oar: Readings on Courting and Marrying*, ed. Amy A. Kass and Leon R. Kass (Notre Dame, IN: University of Notre Dame Press, 2000), 565. The citation is slightly paraphrased from Shakespeare's *Julius Caesar*, I.ii.94–95.

5 Arthur Caplan, "It's not immoral to want to be immortal," MSNBC.com, April 25, 2008. http://www.msnbc.msn.com/id/23562623.

6 Leon R. Kass, *Life, Liberty, and the Defense of Dignity: The Challenge for Bioethics* (San Francisco: Encounter Books, 2002), 257–274.

7 *Ibid.*, 262.

8 Tellingly, however, Kass grants that there might be certain activities—a desire to learn and know, or the experience of friendship, are his examples—that engage our powers without needing "finitude as a spur" (267). He characterizes these as "rare exceptions," but one wonders whether that can possibly be the right way to describe so pervasive and central a human activity as friendship.

9 I wonder, though, whether the following sentence is not in danger of losing Kass's own vision of human beings as finite, embodied creatures: "Through moral courage, endurance, greatness of soul, generosity, devotion to justice—in acts great and small—we rise above our mere creatureliness, spending the precious coinage of the time of our lives for the sake of the noble and the good and the holy" (268). But there is nothing "mere" about our status as creatures who are, on the one hand, attached to particular people and places, and, on the other hand, able to sacrifice even very deep attachments when morality requires us to.

10 Daniel Callahan, *The Troubled Dream of Life: Living with Mortality* (New York: Simon & Schuster, 1993), 138.

11 C. S. Lewis, *The Screwtape Letters, with Screwtape Proposes a Toast* (New York: Macmillan, 1961), 41.

12 Ryan T. Anderson and Christopher Tollefsen, "Biotech Enhancement and Natural Law," *The New Atlantis*, no. 20 (Spring 2008), 99–100. Anderson and Tollefsen also write that it is this "openness of the horizon of goods that makes death an evil" (99). This I find less persuasive. As long as our species survives, as long as there are some of us to realize this open horizon of goodness, it is harder to see why the death of some of us should be an evil. Harder, that is, to see why it should be an evil for humanity. But because each of us has not only human but also personal dignity, because God knows each of us by name, death is an evil.

13 C. S. Lewis, *The Last Battle* (New York: Scholastic, Inc., 1995), 206–207.

14 Psalm 42:1–2.

15 St. Augustine, *Confessions*, 10.40.

16 Kass, 265.

17 The argument in this paragraph draws on J. David Velleman, "A Right of Self-Termination?" *Ethics*, no. 109 (April, 1999), 614–615.

CHAPTER SEVEN · CONFUSIONS

1 See in particular *Human Cloning and Human Dignity* (New York: Public Affairs, 2002), and *Beyond Therapy: Biotechnology and the Pursuit of Happiness* (New York: Regan Books, 2003), especially chapter three.
2 Ruth Macklin, "Dignity is a useless concept," *British Medical Journal*, no. 327 (December 20, 2003), 1419–1420.
3 See Childress's remarks in the transcript at http://www.bioethics.gov/transcripts/dec05/session5.html. See also James F. Childress, "Epilogue: Looking Back to Look Forward," in James F. Childress, Eric M. Meslin, and Harold T. Shapiro (ed.), *Belmont Revisited: Ethical Principles for Research with Human Subjects* (Washington, D.C.: Georgetown University Press, 2005), 246: "[T]he principle of human dignity is central ... [in the Council's report, *Human Cloning and Human Dignity*] and the concept of dehumanization is parasitic on that principle. However, while referred to approximately fifteen times in the report, human dignity is nowhere clearly spelled out. Hence, readers cannot easily determine whether, how, and why human reproductive cloning constitutes dehumanization."
4 Steven Pinker, "The Stupidity of Dignity," *The New Republic* (March 28, 2008), 28–31.
5 *Ibid.*, 28.
6 For just one example of many that might be given, we might point to Albert R. Jonsen, *The Birth of Bioethics* (New York and Oxford: Oxford University Press, 1998). Jonsen discusses at length five different problem areas which were at the heart of the development of bioethics: research ethics, genetics and ethics, the ethics of organ transplantation, issues in death and dying, and the ethics of human reproduction.
7 Pinker, 31.
8 *Ibid.*, 28.
9 F. Daniel Davis, "Human Dignity and Respect for Persons: A Historical Perspective on Public Bioethics," *Human Dignity and Bioethics: Essays Commissioned by the President's Council on Bioethics* (Washington, D.C.: 2008), 26.
10 Pinker, 28.
11 *Ibid.*, 28.
12 Roberto Andorno, "Dignity of the person in the light of international biomedical law," *Medicina e Morale* (2005:1), 92.
13 Pinker, 28.
14 See Mary Ann Glendon, *A World Made New: Eleanor Roosevelt and the Universal Declaration of Human Rights* (New York: Random House, 2001), 217.
15 Pope John XXIII, *Pacem in Terris*, par. 143–144.
16 I note, in passing, how ill-informed is Pinker's animus against religious thought and religious thinkers. Discussing what we might characterize as the internal politics of the President's Council, he refers to an occasion when the appointments

of two members of the Council (Elizabeth Blackburn and William F. May) were not renewed. (Pinker says that "[Leon] Kass fired . . . them," but this is only one inaccuracy among many.) What is of more interest here is that, in his attempt to see the Council's use of dignity language as the result of a kind of religious cabal, Pinker describes William May as a "philosopher" and writes that May and Blackburn (a biologist) were replaced by "Christian-affiliated scholars." I suspect that May himself might be surprised to see himself so characterized. He has been a scholar and teacher of religion for decades and, indeed, was one of two members of the Council who were trained as theologians. Pinker's thesis proceeds untroubled by such factual matters.

17 Pinker, 28.

18 *Ibid.*, 30.

19 *Ibid.*, 29. A peculiarity of Pinker's review is that it spends so much time attacking Kass, even though the collection under review was published after Kass had resigned as chairman of the Council and been replaced by Edmund Pellegrino.

20 See Leon R. Kass, *The Hungry Soul: Eating and the Perfecting of Our Nature* (New York: The Free Press, 1994), 148–149.

21 Kiddushin 40b, as quoted in Kass, 129.

22 The other two are also misleading, but in less significant ways. Pinker claims that dignity is relative and is harmful. But his discussion of its relativity is, in the first place, concerned only with what I have called human dignity and not at all with personal dignity. More important still, giving examples to show that people have sometimes disagreed about what is or is not dignified hardly settles the normative question about the true meaning of human dignity. People have always disagreed about normative claims; that hardly amounts to an argument that such claims are relative. Likewise, his argument that dignity is sometimes harmful again confuses the descriptive with the normative.

23 Pinker, 30.

24 Richard Stith, "The Priority of Respect: How Our Common Humanity Can Ground Our Individual Dignity," *International Philosophical Quarterly*, vol. 44, no. 2 (June, 2004), 170–171.

25 *Ibid.*, 180.

CHAPTER EIGHT · EQUAL PERSONS

1 Kurt Bayertz, "Human Dignity: Philosophical Origin and Scientific Erosion of an Idea," in Kurt Bayertz (ed.), *Sanctity of Life and Human Dignity* (Dordrecht, Boston, London: Kluwer, 1996), 73.

2 Wilfred M. McClay, "God Rest Ye Merry," *Touchstone*, vol. 19, no. 10 (December 2006), 3. Italics added.

3 David Cairns, *The Image of God in Man* (New York: Philosophical Library, 1953), 10.

4 Mary Ann Glendon, *A World Made New: Eleanor Roosevelt and the Universal Declaration of Human Rights* (New York: Random House, 2001), xix.
5 Jacques Maritain, "Introduction," in *Human Rights: Comments and Interpretations.* A Symposium edited by UNESCO (New York: Columbia University Press, 1949), 9.
6 C. S. Lewis, *The Abolition of Man* (New York: Macmillan, 1947), 92.
7 Glendon, 232.
8 St. Augustine, *City of God*, 12.22.
9 David Novak, "Theology, Politics, and Abraham Joshua Heschel," *First Things*, no. 180 (February 2008), 29.
10 Steve Kloehn and Paul Slopek, "Humanity still at heart, soul of cloning issue," *Chicago Tribune* (March 2, 1997).
11 Walter Berns, *For Capital Punishment* (Lanham, New York, & London: University Press of America, 1991), 163. One might contrast this with Robert Merrihew Adams, *A Theory of Virtue: Excellence in Being for the Good* (Oxford: Clarendon Press, 2006), 26: "Moral virtue is a truly worthy object of aspiration; it is right to want it very much. But it is deeply wrong to suppose that any degree of moral virtue could make one 'worth more,' or morally more important, than other people."
12 Robert Spaemann, *Persons: The Difference between 'Someone' and 'Something'* (Oxford: Oxford University Press, 2006), 4.
13 *Ibid.*, 185.
14 Søren Kierkegaard, *Works of Love* (Princeton, NJ: Princeton University Press, 1995), 158.
15 *Ibid.*, 342.
16 Oliver O'Donovan, *The Ways of Judgment* (Grand Rapids, MI: Eerdmans, 2005), 40.
17 J. David Velleman, "A Right of Self-Termination?" *Ethics*, no. 109 (April 1999), 615.
18 Velleman, 615.
19 Gabriel Marcel, *The Existential Background of Human Dignity* (Cambridge, Massachusetts: Harvard University Press, 1963), 134.
20 Marcel, 134.
21 In what follows immediately, I am relying in considerable measure on O'Donovan, 42–49.
22 Kierkegaard, 89.
23 O'Donovan, 44.
24 Eugen Rosenstock-Huessy, *The Driving Power of Western Civilization* (Boston: Beacon Press, 1950), 55–56. Cited in Marguerite Shuster, *The Fall and Original Sin* (Grand Rapids, MI: Eerdmans, 2004), 170.
25 Kierkegaard, 51–52.
26 O'Donovan, 45.
27 *Ibid.*, 48.

28 A wonderful phrase, which I owe to Russell Hittinger. See also Luke 17:7–10.

29 *Taking Care: Ethical Caregiving in Our Aging Society.* A Report of the President's Council on Bioethics (Washington, D.C.: September 2005), 106f. Also available at http://www.bioethics.gov/reports/taking_care/index.html

30 *Ibid.*, 106.

31 *Ibid.*, 104.

32 *Ibid.*

33 *Ibid.*, 107.

34 Ralph McInerny, *I Alone Have Escaped To Tell You: My Life and Pastimes* (Notre Dame, IN: University of Notre Dame Press, 2006), 162.

35 Spaemann, 241.

INDEX

A NOTE ON THE TYPE

Neither Beast nor God has been set in Minion, a type designed by Robert Slimbach in 1990. An offshoot of the designer's researches during the development of Adobe Garamond, Minion hybridized the characteristics of numerous Renaissance sources into a single calligraphic hand. Unlike many early faces developed exclusively for digital typesetting, drawings for Minion were transferred to the computer early in the design phase, preserving much of the freshness of the original concept. Conceived with an eye toward overall harmony, Minion's capitals, lowercase letters, and numerals were carefully balanced to maintain a well-groomed "family" appearance—both between roman and italic and across the full range of weights. A decidedly contemporary face, Minion makes free use of the qualities Slimbach found most appealing in the types of the fifteenth and sixteenth centuries. Crisp drawing and a narrow set width make Minion an economical and easygoing book type, and even its name evokes its adaptable, affable, and almost self-effacing nature, referring as it does to a small size of type, a faithful or favored servant, and a kind of peach.

SERIES DESIGN BY CARL W. SCARBROUGH